Building the Pyramids

How did they do it?

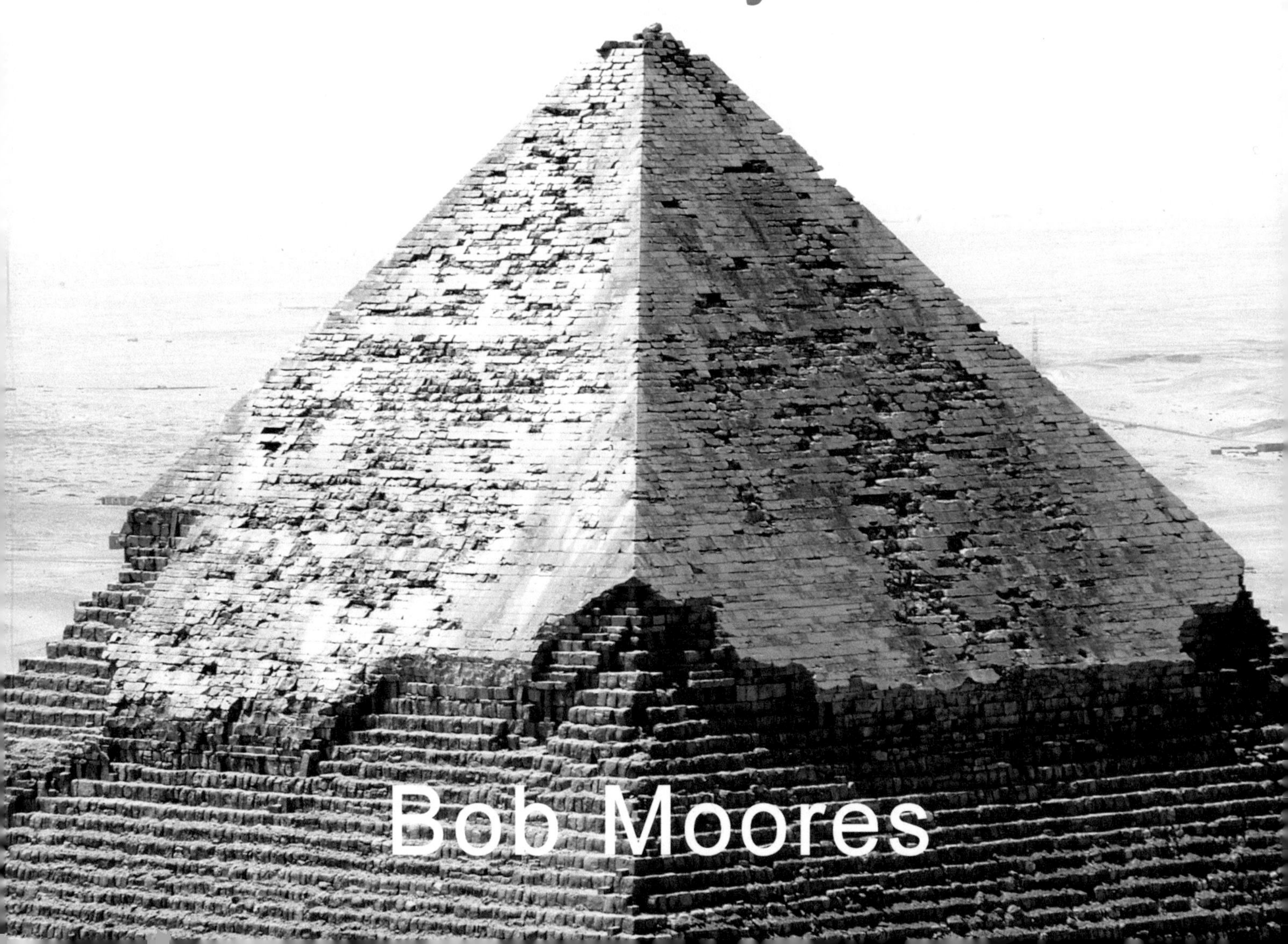

Bob Moores

Building the Pyramids
How did they do it?

iUniverse books may be ordered through booksellers or by contacting:

iUniverse
1663 Liberty Drive
Bloomington, IN 47403
www.iuniverse.com
1-800-Authors (1-800-288-4677)

ISBN: 978-1-5320-7704-3 (sc)
ISBN: 978-1-5320-7705-0 (e)

Library of Congress Control Number: 2019907587

Print information available on the last page.

iUniverse rev. date: 06/20/2019

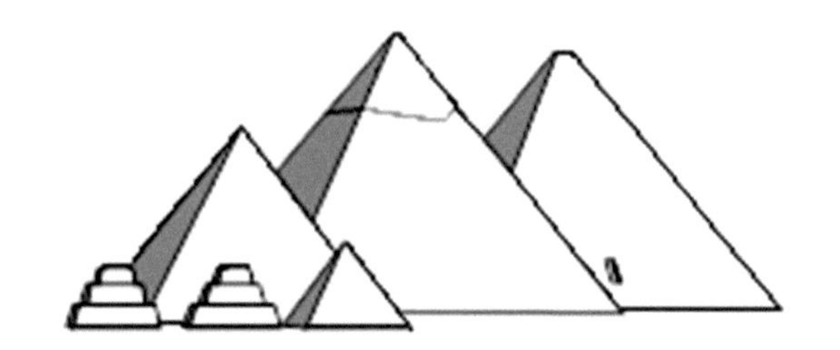

Building the **Pyramids**

How did they do it?

CONTENTS

INTRODUCTION

I'm standing beside the Great Pyramid of Khufu at Giza, Egypt. It's just after sunrise on March 23, 1978. I've dreamed of this moment for so long I was afraid it might be anticlimactic.

Fig. 1 The northeast angle of the Great Pyramid.

I worried needlessly.

Before me a mountain of stone blocks climbs to the sky.

I'm awestruck, mouth agape.

Why? Is it the size of the mountain? Yes, that's part of it. Its age? That too. Its intriguing history? No doubt.

But gazing upward a single question prevails: How in blazes did all those stones get from down here to up there? It is remarkable that we have no answer, though it has been studied for centuries.

Now my baby brother and I are scrambling up the northeast edge of the pyramid. To slow his pace I frequently pause to take photos; otherwise he'll leave me far behind. We gain the summit in twenty minutes.

The view of Khafre's pyramid is breathtaking. The morning sun reflects from its smoothed upper portion. What a sight the pyramids must have been before their beautiful casing stones were plundered long ago.

Fig. 2 Khafre's pyramid, from the top of Khufu's.

Old questions are reborn: How long did it take to build a pyramid? How many people labored in the effort? What were they like? What kinds of machines or tools did they use? How could they build so well? If I had a time machine my first stop would be here, 4500 years ago. Without it, my only option is to discover and interpret clues.

In my teens I became intrigued by ancient construction, human-fashioned works unaided by mechanical engines. The epitome of monumental building was the work of the Egyptians. Whenever a new theory or discovery was announced I would devour every morsel. Sometimes I questioned the author's interpretation or ideas, but was ill-equipped to argue.

Then, in 1973 came two bolts from the blue, Peter Tompkins' *Secrets of the Great Pyramid* and Erich von Daniken's *Chariots of the Gods.* Tompkins presented fascinating description and history of the Great Pyramid, along with theories on its purpose. Most of the latter pointed to reasons other than funerary: it was an observatory, bureau of standards, surveyor's marker, etc. Von Daniken declared that the pyramids were not erected by Egyptians at all, but by visitors from space! That was the last straw.

I launched a new hobby. Who built the pyramids, how were they built, and for what purpose? I started pyramid research in earnest. A few weeks later I had the good fortune to meet Hans Goedicke, then Chairman of Near Eastern Studies at Johns Hopkins University. Dr. Goedicke was most supportive of my research, which he called "homework." For the next five years I spent many Saturday mornings taking notes three stories below ground in the Egyptology stacks of the Eisenhower Library on the JHU campus. Anything related to pyramid construction. Exhausting the books, I was ready for Egypt. I was 39. I asked my brother if he would like to be my accomplice. Greg, twenty and an engineering student at the University of Maryland, jumped at the offer. He later joked that I only wanted him as a "bearer," since *his* backpack somehow attracted most of our stone samples.

We spent two memorable weeks in Egypt in the spring of 1978 (don't eat the ice cream). We took 1800 color-slide photos, some of local color, but most of pyramid details. I annotated many with a mini-cassette (tape) recorder. Both technologies now seem ancient. We retrieved nearly two hundred rock samples. The Customs Inspector at Cairo Airport didn't seem to regard our bounty as unusual: crazy Americans coming to Egypt to get rocks. *Please take all you want.*

Our lone disappointment was being denied access to the Dahshur pyramids. These were close to a military base, and our previously approved security clearances had become lost in Egyptian red tape.

That glitch was remedied in 1987. I returned as part of the National Geographic team that revealed the second of Khufu's two wooden ships interred in a rock-cut pit on the south side of his pyramid. I was responsible for drilling a hole through a five-foot-thick monolith of the pit ceiling. A special airlock I designed allowed air sampling and imaging of the pit interior while preventing transfer of air into or out of the pit. My daughter Sheri, a junior at the University of Virginia, was thrilled to be hired by Nat Geo as a backup photographer.

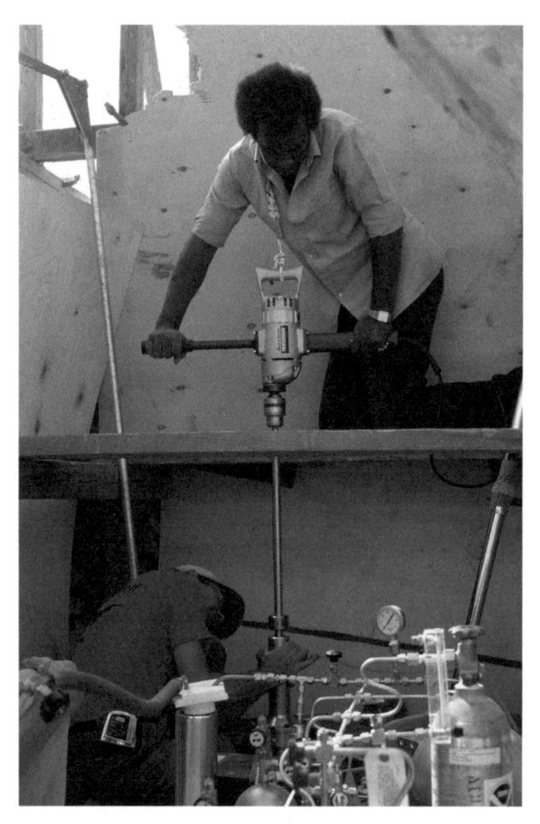
Fig. 3 My friend Haggag Hassan runs the drill while I check its progress. Photo by Sheri Moores.

The project allowed time for more research. Because of my newly elevated standing with Egyptian officialdom, Sheri and I were allowed to visit the Dahshur pyramids. I also received permission to explore and photograph the enigmatic Well and Grotto under the Great Pyramid.

My broader objectives condensed to a single ambition: **Discover the *most probable* way the pyramid stones were raised.** My fallback goal became: **Find a *feasible* means by which the stones were raised.**

The pyramids are not just huge piles of stone. Each is unique. In them we can trace a succession of architectural advances. People who study the pyramids differ on their purpose and means of construction. I'll analyze these opinions, adding my own, but never under the illusion that I am *proving* anything.

The data I gathered ranges from soft to hard. Soft data are observations and speculations from explorers ancient and modern. Hard data are more tangible, physical evidence we see now: writings, paintings, sculptures, tomb furnishings, and of course the pyramids themselves.

Now for a few bookkeeping items.

This work is focused on pyramid building. It is not a book of Egyptian history. It touches history and religion only for context in pyramid evolution.

I provide observations that relate to construction methods. You can find detailed measurements and descriptions elsewhere. I particularly recommend *The Architecture of the Memphite Pyramids,* by Italian scholars Vito Maragioglio (1915-1976) and Celeste Rinaldi (1902-1977). They explored the Old Kingdom pyramids in the late 1950's and early '60's. In seven volumes they give precise measurements and wonderful drawings. They also present logical analyses of what they found.

The measurements I *do* give are in metric units followed by English equivalents, with one exception: Weights of blocks given in "tons" refer to U.S. tons of 2,000 pounds. The few engineering calculations are also in English units because that's how I was schooled.

The builders employed neither system. For length measurements they used a *cubit* of seven *hands*, each of four *fingers.* The precise length of the cubit is unknown, but we can infer a value by measuring Old Kingdom buildings. For example, Cole (1925) found that the average length of the four sides of Khufu's pyramid is 230.364 meters. Divide that by the 440 cubits that Egyptologists believe Khufu's builders had set, and we get a cubit of 52.355 cm (20.61 inches).

The photos and drawings in this book, unless otherwise noted, are mine. In some of my close-up photos you'll see an engineer's scale. I placed it so the size of the subject can be judged. The scale has sixteen centimeters painted alternately black and white. For large-scale reference I sometimes include Greg in the picture. He is 170 cm (five feet, seven inches) tall.

For other measurements I designed a device (seen in fig. 149) to which I attached an angle-measuring component, or *clinometer.* One edge of the gadget is a cubit long, the other a half-meter. Both edges were milled straight to within 0.127mm (.005 inch). The clinometer is accurate to about plus-or-minus ten minutes of arc. I also carried a steel measuring tape.

This book is organized in what I hope is a logical sequence. Chapter one is about the people who built the pyramids. Chapter two addresses materials of the pyramids, from where and how they were obtained. Chapter three is on methods used to transport blocks to the pyramids. The next four chapters encompass the seven early pyramids that inspired this book. I address them in the order that *professional* Egyptologists believe they were built. Chapter eight is a glance at later pyramids. These have a few interesting features, but are generally smaller, more cheaply made, and more ruined. Chapter nine looks at masonry techniques the builders employed or may have employed. Much of that is conjecture, but not as speculative as how the stones were raised, the subject of chapter ten. You can think of chapters one through nine as compiling evidence for the discussion in chapter ten.

In chapter ten I examine stone raising theories. I will be quite critical, in some cases harshly so. I'll explain why. Some of these ideas could be partly right, even my own, but I hope to be objective in presenting the good and not-so-good points of all.

In this book *pyramid builders* is an appellation that includes all who ordered, planned, supervised, and worked on these amazing structures. From mortar mixer to architect every job was necessary. Because responsibility was so-shared, I credit them all.

A personal proclivity: If you noticed a few paragraphs back, I mentioned professional Egyptologists with emphasis on "professional." I suppose there are more *amateurs* (I am one) than professionals. The line between amateur and professional may be a little blurry, but professionals are doing the actual work. Professionals pretty much agree on the timeline and sequence in which the pyramids were built. Many amateurs disagree. They often propose much greater ages (by thousands of years), particularly for the Great Pyramid. For reasons that will become apparent, I follow guidance of the pros.

Some books and articles on the pyramids have titles rather pretentious and misleading. They boldly imply that the mystery of pyramid building has been solved, but in the text you find nothing of the sort. Truth is, we are guessing. No one knows the exact methods the builders used to create their architectural wonders. But three things we can confidently say: The builders were superbly organized, brilliantly innovative, and craftsmen *par excellence.*

Masonry and pyramid terms

Angle (of the pyramid)	–	A corner of the pyramid.
Architrave	–	A large stone beam that forms the ceiling of a passageway.
Apothem	–	On a triangular pyramid face, the midline that is perpendicular to its base.
Backing stones	–	The blocks immediately behind (in back of) the casing stones.
Batter	–	The slightly backward (off-vertical) slope of masonry courses in a wall.
Bedding joint	–	The joint between the bottom of a block and the block it sits on.

Bedding seat – The surface upon which a block of the next higher course sits.

Casing stones (or **Facing stones**) – The blocks forming the outer face of the pyramid.

Closure block – The last block placed in the middle of a row of blocks.

Corbelling – Successive offsetting of block courses, narrowing an opening, as courses rise.

Causeway – A straight, paved road between valley and pyramid temples.

Courses – Horizontal rows of blocks in a masonry wall or pyramid face.

Dressing – Trimming or smoothing the face of a block.

Finish-dressed block – A block with its outer face trimmed and smoothed flat.

Header – A rectangular block with its long dimension perpendicular to a wall or pyramid face.

Hip lines – The four lines representing the corner edges of the regular pyramid form.

Joint – The junction between adjacent blocks, sometimes filled with mortar.

Mastaba – A tomb having a low rectangular (boxlike) superstructure.

Mortar – Cement placed between adjacent blocks to help bind them together.

Nucleus – A structure within the body of a pyramid, often in the form of large steps.

Packing stones – The blocks, not well-squared, which lie behind the casing and backing stones.

Portcullis – A stone slab lowered vertically or obliquely to block a passageway.

Pyramidion – The topmost block on a pyramid, the capstone in the form of a small pyramid.

Rising joint – The joint that rises vertically or nearly so between adjacent blocks.

Sarcophagus – In a pyramid, a stone coffin.

Seked – Ancient Egyptian term for the slope of a pyramid face. Expressed as the horizontal component of a slope having a vertical component of seven. Example: A seked of five is a slope of five horizontal units for every seven vertical units.

Sleepers – Wooden beams placed in soft ground to prevent sled runners from sinking in.

Step pyramid – A pyramid in the form of large steps.

Stretcher – A rectangular block with its long dimension parallel to a wall or pyramid face.

True (or regular) pyramid – A pyramid that has four flat, identical, triangular faces.

Voussoir – A wedge-shaped stone block that forms part of an arched ceiling.

Zero level – The surface of the pavement upon which the first course of pyramid blocks rest.

1

THE PYRAMID BUILDERS

Seventeen large pyramids are found in a 72 km (45 mile) stretch of desert on the west side of the Nile between Abu Roash and Meidum. Several were abandoned before completion. Small pyramids adjoin the main pyramids, serving either as queens' tombs or for cult worship. Many smaller pyramids are found southward to central Sudan, but these are not the subject of my study.

Fig. 4 Pyramids and quarries in the Nile valley.

In this book I focus on the seven largest, best-preserved, first-completed pyramids. They were raised during what modern historians call the "Old Kingdom," and appear to be composed of stone masonry throughout. The seven are drawn to the same scale and numbered in order of construction in figure 5. Spanning the reigns of Djoser in Dynasty 3 to Menkaure in Dynasty 4, a period of about 150 years, they contain perhaps twenty-four million tons of limestone. The *dynastic* nomenclature of related family groups, incidentally, was created by Egyptian historian Manetho, who lived during the Greek occupation of Egypt in the third century BCE.

Egyptians built the first pyramid during the reign of Djoser (2630-2611 BCE), about four hundred years after the unification of the "two lands" of Upper and Lower Egypt. Upper Egypt, being up-river (the Nile), refers to southern Egypt; Lower Egypt is down-river to the north.

The builders' society was structured, fittingly, as a pyramidal monarchy. At the apex was the god-king. He was not simply a figurehead. On ascendency to the throne he became the latest incarnation of Horus, the falcon god of the living and the sky. The king had up to five names, depending on the period in which he lived. In addition to his Horus name, he could be identified by the combination of his *nomen* and *prenomen*, these names encircled by ovals (cartouches) which signified his rule over the entire world.

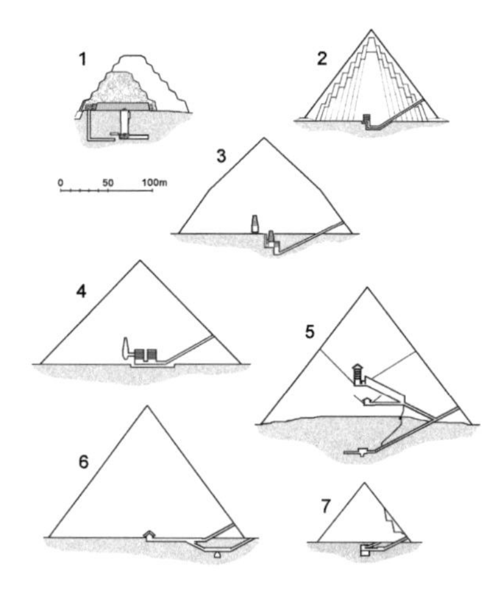

Fig. 5 The stone pyramids of Dynasties 3 and 4. 1) Djoser's Step Pyramid. 2) Sneferu's Meidum Pyramid. 3) Sneferu's Bent Pyramid. 4) Sneferu's Red Pyramid. 5) Khufu's Great Pyramid. 6) Khafre's Pyramid. 7) Menkaure's Pyramid.

Members of his royal family served as advisers and transmitters of his orders. Nepotism prevailed. Second only to the king, a close male relative, often a nephew, was appointed vizier (va-ZEER). His official title was "overseer of all the king's works." Of the king's works, none were more important than his tomb and temples. In the temples, rituals were performed by a retinue of priests to propitiate the gods, and thus ensure the king's comfortable, eternal afterlife. The priesthood was the repository of arcane knowledge, encompassing the combined functions of Science, Engineering, *and* Religion.

Each of the forty-two provinces of Egypt was administered by a governor.

After relatives, priests, and governors came officials of every *raison d'être*. In 1938-39, while excavating the mastaba cemeteries east and west of Khufu's pyramid, Egyptian archaeologist Selim Hassan (1887-1961) found several hundred titles (1960, X, 53) that tell us much about the VIPs. A few examples:

The Honored by the King of Upper and Lower Egypt
Chief of the Dancers of the King
Assistant Waiter of the Cooler of Drinks
Great Seer (High Priest of Heliopolis)
Scribe of the Gangs of Workmen
Inspector of the Scribes of the Gangs of Workmen
President of the Shipyard
Captain of the Two Divine Boats
The One Concerned with the King's Affairs
Master of Secrets
Mistress of Pleasures
Director of the Dwarf's Linen
Scribe of the Four Divisions of Artisans

Overseers of: the Gangs of Workmen, the Workshops, the Hairdressers of the palace, the Embalmers, the Treasury, the Canals, the Game Preserve, the Great Court of Justice, the Artisans of the King's Ornaments, the Physicians, the Fowling Marshes, the Tutors of the King's Children, the Service of Food, the King's Harem.

My favorite: He Who Does What His Lord Likes Every Day (substitute "Wife" for "Lord" and you have my job title).

Skilled people worked on royal projects year-round. Block-laying masons, sled makers, rope makers, quarrymen, ramp builders, sculptors, and boat builders were among the hundreds of jobs required to produce the pyramids, tombs, temples, and other government works. In addition to transporting building stones via the Nile, sea-going merchant vessels brought highly valued materials like cedar from Lebanon and copper from the Sinai.

I find interesting the importance of scribes, the record-keepers and accountants of all aspects of business and labor. The Egyptians did not use currency. They traded services and commodities, an example being measures of grain. When I worked in product development we poked fun at our accountants by calling them "bean-counters." But in ancient Egypt they actually were. Scribes were adept at manipulating fractions, as payments had to be allocated to groups or individuals according to their exact contributions. They even recorded which group of masons provided a particular building block and its dimensions.

Egyptologists believe that a large portion of pyramid workers were conscripted, a form of taxation, especially during the 3-4 months every summer when the Nile flooded most of the farmland in the Valley. During that period many farmers would have been available for government work. High Nile was also the best time to ship casing stones from the quarries on the east side of the river, as cargo boats could have been brought close to the pyramids.

As for the stone haulers, American archaeologist George Reisner (1867-1942), in excavating the mortuary temple of Menkaure (Reisner used his Greek name, *Mycerinus*) in 1906-1907, found builders' marks on many of the core blocks. He reported (1931, 276) that the stone-cutters-haulers-setters were organized in two *crews* of 800-1000 men each. One of the crews was named *Mycerinus-is-drunk*, the other *Mycerinus-excites-love*. Though he found the names of only two crews in Menkaure's temple, he said four have been found for Khufu's pyramid.

The crews were composed of five 200-man groups which Reisner translated as *watches,* but are now called *phyles*. The five phyles of both crews were named using the same nautical terms: *Bow, Stern, Port, Starboard*, and (per Lehner) *Last*. All the core blocks on which these marks appeared were of local limestone, and so had not been transported over water. Said Reisner, "The nomenclature of the boat watches was transferred to the watches or companies of priests and workmen."

Reisner thought that because many of the blocks which contained watch names also had "distinguishing marks" (e.g. a bird or animal), these marks probably indicated that the watches were subdivided into working gangs of 10-50 men.

Reisner also noted that the term "'desert workshop' (or similar), is a constant element occurring on every block on which the inscription is legible." This may indicate that the blocks, after being quarried, were brought to a building yard to be trimmed-to-order for their final positions in the temple.

Returning to watch names, perhaps these were related to the positions and duties of each group. The port and starboard watches may have corresponded to the left and right row of sled haulers. The stern team may have been those who cut blocks free in the quarry, the bow team those who unloaded the blocks, and the last team those who set the blocks in position. I am guessing here.

Of course there would have been people who supplied the builders with necessities like food, water, clothing, and shelter. In other words, pyramid building would have required the infrastructure of a large town.

Hassan thought that the pyramid city of the Giza builders would be found nearby, but did not command resources to locate it in 1939. That would change by the end of the century.

In the 1990's much was discovered about the builders, principally in excavations led by Zahi Hawass, Director General of the Giza pyramids, and American Egyptologist Mark Lehner. Though the Giza pyramids were not the first to be built, it is here that we learn most about the builders.

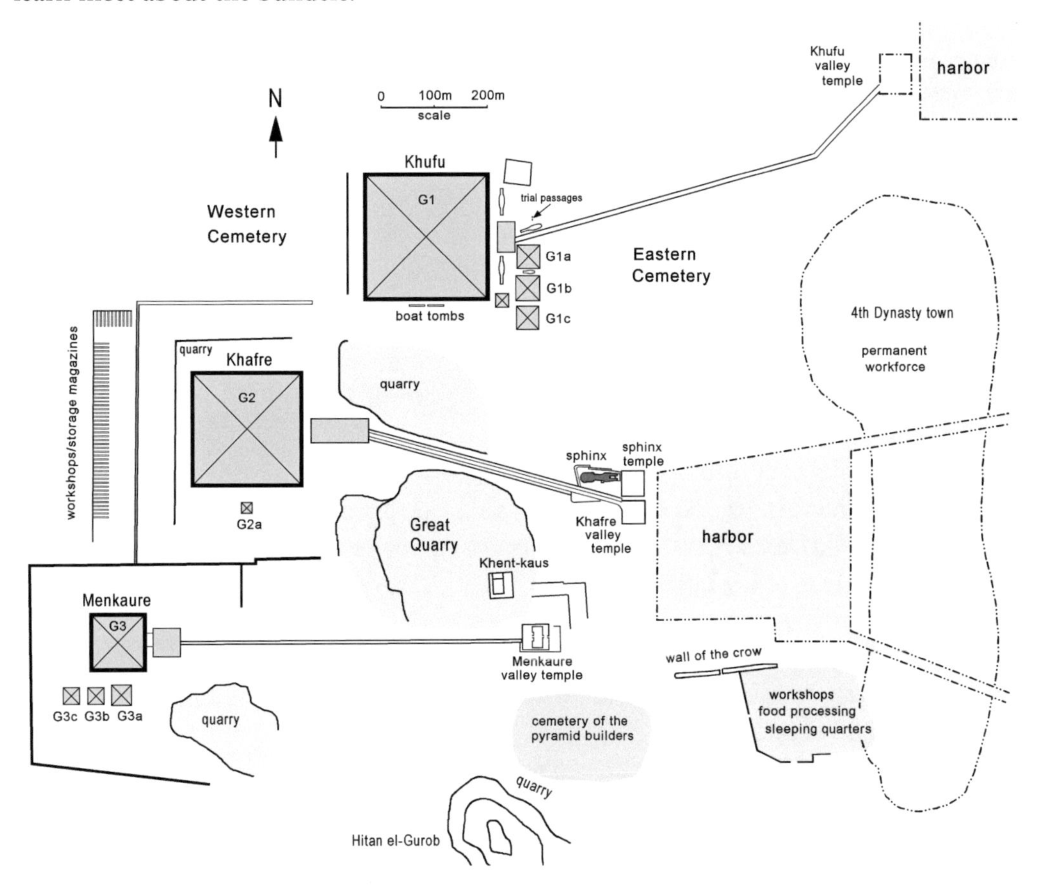

Fig. 6 Map of the Giza pyramids. Redrawn from Hawass (2011). The dotted lines reflect areas of limited exploration.

In 1990 Hawass and Lehner discovered what looked to be a food production center for the workers. It is located just south of a large east-west wall called the "Wall of the Crow." A more extensive excavation began in 1995, outlining the east and south borders of a larger complex. Craft shops, bakeries, beer breweries, a fish processing center, and long galleries that could have been sleeping quarters were revealed. North of the compound, extending right up to the valley and sphinx temples of Khafre, Hawass found the quay of a large harbor. Stones shipped from Aswan and the quarries east of the Nile were almost certainly delivered here.

East of the harbor a large town has been delineated by discoveries, some accidental, of ancient residences. But since the modern village of Naslet el-Samman now covers that area, further excavation is problematic. Nevertheless, Hawass and Lehner believed they found part of an enclosure of the royal residence of the pyramid kings. They called this their "palace hypothesis."

West of the workshops and food processing area, and south of Menkaure's valley temple, Hawass found tombs of the pyramid builders. He reported in *Archaeology* vol. 50-1 (1997):

"The cemetery of the laborers and artisans south of the Wall of the Crow at Giza reveals as much about life and death in the Old Kingdom as do the burials of royalty and noblemen north of the wall and surrounding the pyramids. A lower part of the burial ground contains small mud-brick and stone-rubble tombs of common laborers, while a higher tier is dominated by larger, decorated stone tombs of craftsmen and overseers."

The physical remains of the workers tell us about the hard lives they lived. Hawass, in the same article:

"Many of the men died between the age of 30 and 35. Below the age of 30 a higher mortality was found in females than in males, a statistic undoubtedly reflecting the hazards of childbirth. Skeletons from the great mastaba cemetery west of the Khufu pyramid, in which members of the upper class were buried, reflect a healthier population whose women lived five to ten years longer than those of the artisan and worker community. Degenerative arthritis occurred in the vertebral column, particularly in the lumbar region, and in the knees. . . . Skeletons of both men and women, particularly in the lower burials, show such signs of heavy labor."

Simple and multiple limb fractures were found in skeletons from both the lower and upper burials. The most frequent were fractures of the ulna and radius, the bones of the upper arm and the fibula, the more delicate of the two lower leg bones. Most of the fractures had healed completely, with good alignment of the bone, indicating that the fractures had been set with a splint.

No cause-of-death analyses of the workers' remains were given.

Hawass finishes with refutation of the most popular myth regarding the builders: "The pyramid builders were not slaves, but peasants conscripted on a rotating basis, working under the supervision of skilled artisans and craftsmen, who not only built the pyramid complexes for their kings and nobility, but also designed and constructed their own, more modest tombs."

The moral ideals of the ancient Egyptians

The *Book of the Dead* is a collection of religious texts that date to the New Kingdom, about a thousand years after the pyramid age. They provide insight to the values of the people. Here we find the "negative confession" whereby the supplicant listed all the bad deeds that he never did. Hawass found similar statements in the tombs of the builders at Giza. Notwithstanding the question of truthfulness, we can see what was deemed good and bad behavior, and those notions are not much different than those we hold today.

Another indicator of their moral system is the ritual performed by their gods called "weighing of the heart." The heart was considered the locus of the soul.

Fig. 7 Weighing of the heart. Papyrus of Hunefer, c. 1375 BCE.

In this standard scene, the heart of the deceased was put on a balance scale and weighed against the feather of Maat, goddess of truth and justice. This was one of the tests in the deceased's journey through *duat* (the underworld, and kingdom of Osiris) toward a good afterlife. If his heart was as light as the feather, the deceased was allowed to proceed. If heavier than the feather, the heart was devoured by the tri-bodied (croc/lion/hippo) goddess *Ammit*, relegating the deceased's soul to non-existence. Even today we employ metaphors assigning spiritual aspects to the heart, an obvious referral to this ancient ceremony being the expression "her heart was as light as a feather."

Before leaving the subject of the spiritual I should mention that there are differing interpretations of the ancient Egyptian concepts of the *ka* and *ba*. The simplest view, at least for me, is that the ka was the external representation of a person's being, how a person appeared to the outside world. The ba was the internal soul known only to the individual and his god.

The Egyptians made significant gains in building technology. They developed stone masonry with inspiration, experimentation, and persistence. They did good work. Their products are quite impressive, not just in scale, but also in innovation. Their deeds are even more remarkable when we consider the relatively simple tools and materials they employed.

2

MATERIALS, QUARRYING

The pyramids are made of two types of limestone, a sedimentary rock. The type forming the greater bulk of the pyramids (their cores) is a coarse-grained stone obtained from open-pit quarries close by each building. This is *local* limestone. The pyramids and other important buildings were covered with an outer skin, or casing, of a finer-grained, whiter variety. This stone was taken from underground quarries at Mokattam, Tura, and Maasara on the east side of the Nile (see figure 4), and therefore had to be shipped across the river. For brevity I will refer to this material as *Tura* limestone. Both types are considered soft stones; they can be scratched by your fingernail.

Certain features in the pyramids and adjoining temples are of hard-stone like granite and basalt. These igneous stones cannot be scratched by your fingernail. Though the hard-stone features comprise a tiny percent of total masonry built, they required, for their volume, much more labor to obtain and shape.

I will address in more detail each of these four materials.

Quarrying requires tools, so I will include stone-working tools and methods of block extraction.

Because more archaeological work has been done at Giza than at any other site, my remarks will focus on those pyramids.

Local (core) limestone

Local limestone varies in color, texture, hardness, density, porosity, friability, and fossil content. Differences depend on geographic location and depth in the ground. For example, the plateau on which the Giza pyramids sit is an outcrop of the Middle Eocene Mokattam formation. It is topped with a white, coarse-grained limestone largely comprised of lens-shaped fossils called *nummulites*. Deeper strata, visible in the body of the Sphinx and the walls surrounding it, reveal a finer-grained limestone with fewer fossils apparent. Its colors range from off-white to light tan, yellow, and orange.

Fig. 8 The Great Sphinx.

Today most of the blocks of the outer surfaces of the pyramids are *backing* stones. They were immediately behind the casing blocks that were stripped away centuries ago. Dietrich and Rosemarie Klemm, in *The Stones of the Pyramids _Provenance of the Building Stones of the Old Kingdom Pyramids of Egypt* (2010), say that the backing stones now visible on Khufu's pyramid are the same material as the now-missing casing. They state (p. 88) that "all casing and backing samples [75 samples] from the Khufu pyramid correspond well with these three [Tura, Maasara, and Mokattam] provenance areas." I do not question their samples. But I found many backing stones on the south face of Khufu's pyramid to be the same material as the bedrock on which the pyramid sits: nummulitic limestone. I did not closely examine the backing stones of the other three faces.

Before authorities banned climbing of Khufu's pyramid (about 1980) most climbers ascended the four corners, especially the northeast angle, assisted by pyramid guards who required small fees. The corners are the easiest places to climb because they are not as steep as the faces. The material at these edges and at the top of the pyramid indeed appears to be fine white limestone, perhaps from Tura.

During the 1987 National Geographic project I mentioned earlier, I asked an official of the Egyptian Antiquities Service if I could have a sample of the core masonry of the pyramid. He simply asked "From where?" When I indicated an area to the rear of a second-course backing stone, he took a piece of pipe that was lying nearby and whacked off a piece. Figure 9 shows the sample. Note the heavy concentration of nummulites.

Fig. 9 Core masonry from second-course backing stone, south side, west end of Khufu's pyramid.

Quarrying local limestone

The quarryman's method for extracting a block of limestone may seem labor-intensive, but given his technology it was straightforward. Using tools I'll describe soon, workmen carved a trench to delineate the block to be detached. The trenches for smaller blocks "rarely exceed 4 ½ inches in width," said English Egyptologists Somers Clarke (1841-1926) and Reginald Engelbach (1888-1946) in *Ancient Egyptian Masonry* (1930, 17). Trenches for thicker blocks had to accommodate the body of the quarryman, so these were about forty-five centimeters (1.5 feet) wide. Figure 10 shows trenching and possible lever slots (if ancient) on the west side of the Great Quarry at Giza. In figure 11 you can see trenches where large blocks were taken at the northwest corner of Khafre's pyramid.

Fig. 10 Quarry operations at west boundary of Great Quarry at Giza.

Fig. 11 Quarry at NW corner of Khafre's pyramid.

At the bottom edge of the outer face of the block to be detached, workmen cut notches or a continuous groove that could take the points of wooden levers. The block was then pried upward, parting along a horizontal bedding joint of the layered rock. I think this was the typical practice, but there is no consensus – some think wedges were used.

Fig. 12 Extracting a limestone block.

One theory is that quarrymen placed wooden beams in the continuous grooves, soaked them with water, and waited for the expanding fibers to break the blocks free. Though this technique could possibly work, if actually used I guess the pyramids might still be under construction.

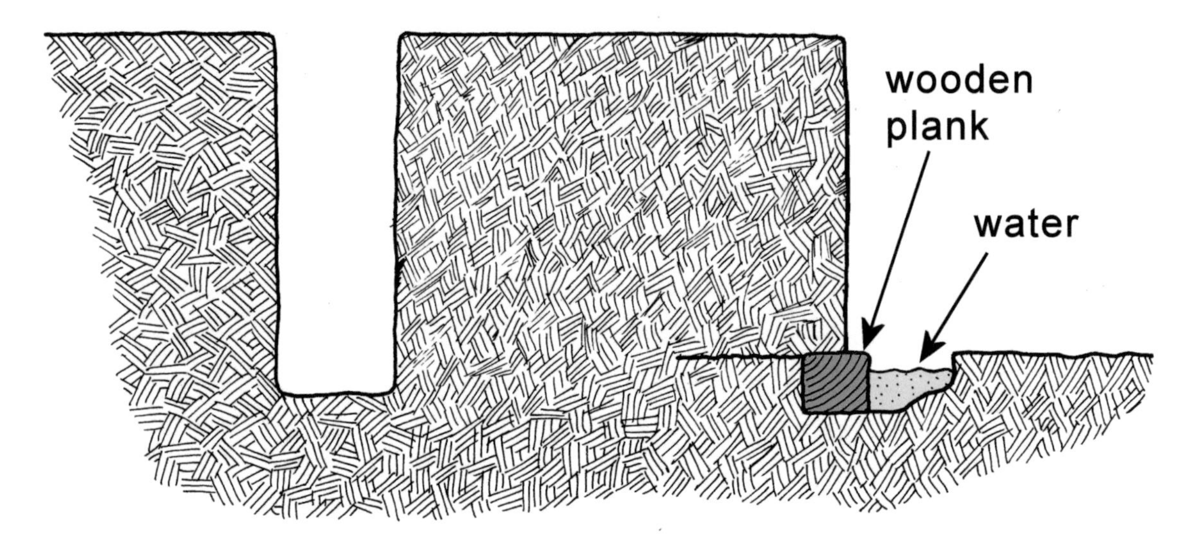

Fig. 13 Freeing blocks using water-swelled wooden beams (unlikely).

Evidence for trenching and levering can be observed in the surface quarries near the Giza pyramids and on their larger blocks. Surprisingly, lever (or wedge) notches are rarely seen on smaller blocks that comprise the bulk of core masonry. To understand this we must look at block extraction in greater detail.

Most of the core blocks are not nicely squared. Although their upper and lower surfaces are reasonably flat, they often have a side that appears broken rather than chisel-cut. It seems that cutting a trench around the entire perimeter was not required for blocks less than three-quarters of a meter (about 2.5 feet) thick.

I propose, without proof, that levers applied from the front, or in a side trench, could separate the block, even with one side attached (or partly attached) to the rock. Again I am speaking of rough core stones less than 75 cm high and at least twice as wide as they are high, the prevalent type. The angular face produced by the splitting-off process apparently was not objectionable to the builders.

With regard to the missing notches, perhaps these incisions, when needed, disappeared in the process of liberating the next lower series of stones.

Tura (casing) limestone

Important buildings were covered with fine white limestone taken from underground quarries on the east side of the Nile. It had to be ferried across the river to harbors close to the pyramids. This handsome stone is fine-grained with no fossils obvious by eyeball. It was formed of countless bodies of marine microorganisms in the Cretaceous period 145 to 66 million years ago. As you can see in figure 14, a broken face of Tura limestone is not pure white, but becomes nearly so when rubbed or polished.

Fig. 14 Tura limestone.

Quarrying Tura limestone

The fine white limestone was quarried by methods similar to those for the local stone. But in this case, workers had to tunnel into cliff faces to obtain it. The tunnels, sometimes exceeding six meters (twenty feet) in height, were cut far into the hillsides. Columns were left at intervals to support the ceiling as the hill was undercut.

Fig. 15 Underground quarry at Tura.

Here quarry faces were worked vertically, top to bottom. At the top of each vertical row of blocks to be separated, a space was excavated horizontally into the rock, allowing access to make the backside trench. The widths of the trenches, as in the open quarries, depended on whether it had to accommodate the workmen's arm or body. Because these operations were carried out in confined spaces it is likely that all four sides had to be exposed before the block could be detached.

Clarke & Engelbach reported that in these "closed quarries" wedges of metal or wood were used to free the blocks from their beds. They cite as evidence "wedge slots" that remain where the lowest courses have been removed. In this case I think wedging is probably right. The fine white limestone is more compact and less stratified (layered) than the local stone, so levering might not work.

Most of the pyramids lost their casing to later building projects, especially during the Middle Ages. But stone pilferage was a late chapter in the violation of these monuments. It's probable the Old Kingdom pyramids were thoroughly looted in the turmoil of the First Intermediate Period, within a few centuries of their building.

Stone-cutting tools

How were the trenches that defined a limestone block made? It is generally accepted that quarrymen chopped and shaped the limestone blocks using copper chisels struck by wooden mallets. Many of these tools have been retrieved, and some have been depicted in tomb paintings.

Fig. 16 Quarryman's mallet and chisel.

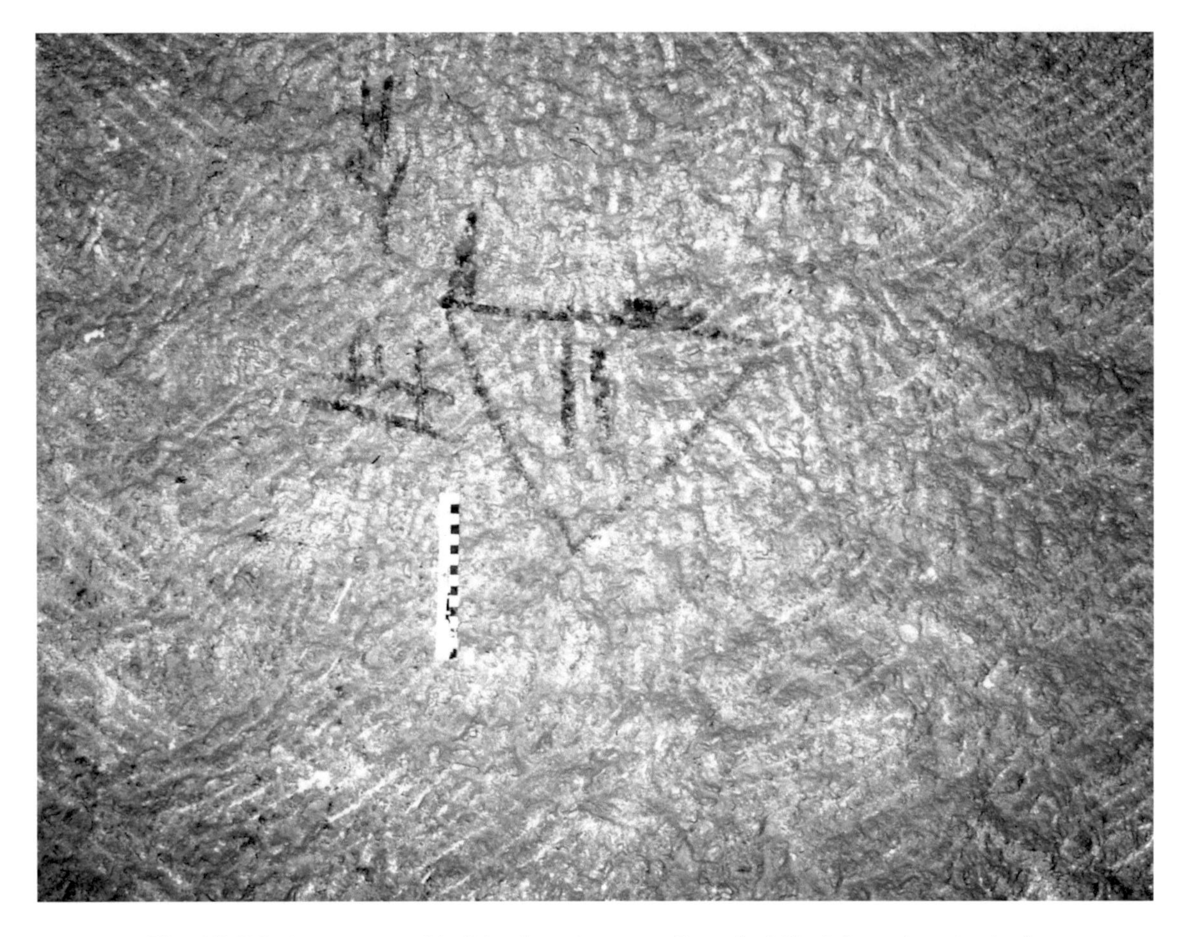

Fig. 17 Chisel grooves and builders' marks on south wall of Khufu's eastern boat pit.

On quarry faces and blocks themselves are grooves which appear to have been cut by a chisel. Further, Zaky Iskander (Nour, et al, 1960) found copper particles in chiseled grooves in the walls of Khufu's eastern boat pit.

Case closed? Not yet.

Virtually all metal implements recovered from Old Kingdom sites are of copper. Further, Old Kingdom copper is of such purity that trace elements rarely exceed two percent of the total. No tools of harder metal, such as bronze or iron, have been irrefutably dated to this period. Thus we are left with the discomforting thought that the masons could cut stone with tools of a metal only slightly harder than the material being worked. To explain why this is bothersome let's make a brief detour into mineralogy.

All stones are composed of minerals, which are compounds or elements having specific chemical formulae. For practical reasons minerals are classified in properties like color, grain size, cleavage, specific gravity, crystallographic structure, and hardness.

Hardness, the resistance to scratching or abrasion, is pertinent here. Early in the nineteenth century German geologist Friedrich Mohs devised a simple scale of mineral hardness that is still in use. Mohs arranged ten minerals such that each will scratch those lower in the scale, but none above. Table 1 shows the Mohs scale along with *relative* hardness values found by the later Knoop indentation test.

Table 1 - Mineral Hardness Scales

Mineral	Mohs	Knoop
Talc (softest)	1	0.5
Gypsum	2	1
Calcite	3	2
Fluorite	4	3
Apatite	5	5
Orthoclase	6	6
Quartz	7	8
Topaz	8	11
Corundum	9	18
Diamond (hardest)	10	70

Glass has a Mohs hardness of about 6.5 to 7.0, hardened steel 5.5 to 6.0, and your fingernail about 2.5. Limestone, whether coarsely fossiliferous or finely crystalline, is almost entirely composed of calcium carbonate, which, in the form of large calcite crystals, has a Mohs hardness of 3.0. In the form of fine-grained limestone, though, calcite is slightly softer. Native copper has a Mohs hardness of 2.5 to 3.0.

Most trace elements in Egyptian copper make the metal harder. Iron, manganese, nickel, tin, bismuth, and arsenic sometimes appear. But we must judge these impurities rather than planned alloying agents because of their paucity and variation.

The hardness of copper can be further increased by hammering, or work hardening. This process was, as British chemist Alfred Lucas (1867-1945) stated (1962, 213), "the only secret of hardening the ancient Egyptians knew," and further "the 'lost art' [of hardening copper] so often referred to is a myth." Lucas then relates an experiment by a Mr. Desch, who hammered copper to the hardness of soft steel (Mohs 4.0). We are left with the question: Can hard copper or soft steel cut limestone?

I made a wooden mallet and hammer-hardened copper chisel. With these I have carved several varieties of Giza limestone. The chisel does not "cut" the stone. In fact, a sharp cutting edge is not needed. The particles of local stone are not tightly bonded. Upon impact the chisel penetrates by crushing the material beneath its point. The blow fractures surrounding material, making small chunks fly off.

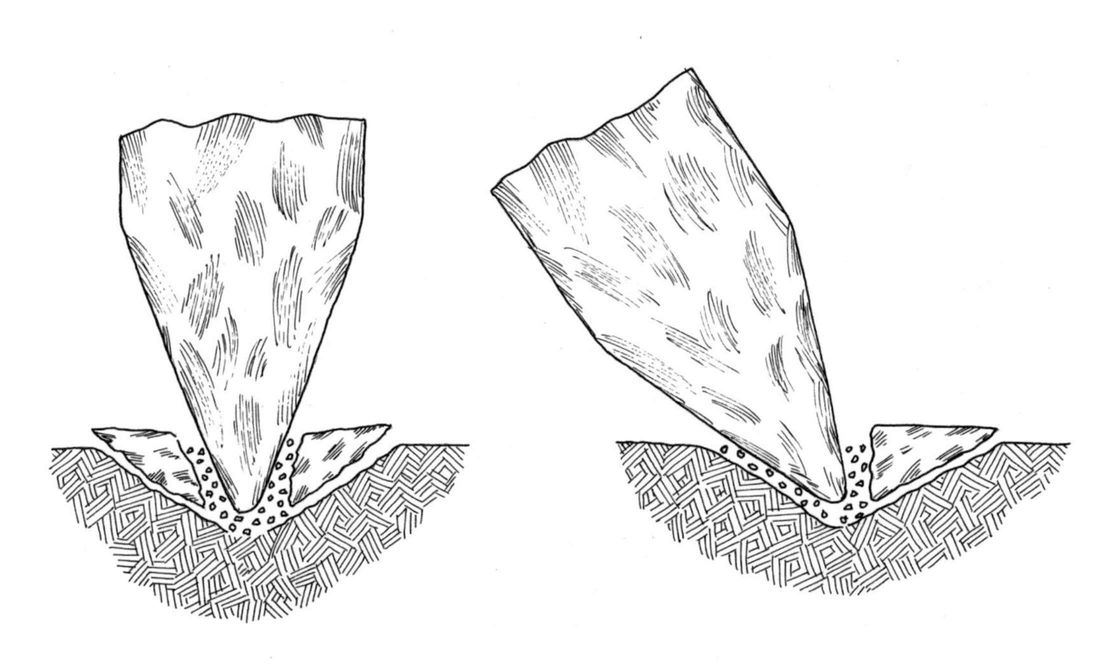

Fig. 18 Chisel action on local limestone.

Tura limestone is a bit different. Though slightly less dense than the Giza stone, it is more compact and less easily chipped. In places where this stone was worked one can notice two types of marks. The first are long, roughly parallel channels that appear to have been gouged out by a pointed tool, perhaps the same that worked the local stone. The second type looks to have been made by a sharp-edged straight chisel blade about 15-18 millimeters (0.6-0.7 in.) wide. The latter form tracks as long as twenty centimeters (8 inches) with no apparent dulling of the edge that produced them.

Fig. 19 Chisel tracks on first-course backing stone of Khufu's pyramid.

Figure 19 reveals chisel tracks on the vertical face of a first-course backing stone at the east end of the south side of Khufu's pyramid. The stone is either Tura limestone or a fine-grained local stone. Fossils are not apparent. The tracks have a saw-tooth form, with cuts spaced about three to five millimeters apart. These tracks were certainly produced by a chisel edge that was held in contact with the work and struck repeatedly, but we cannot be sure that these tracks were made anciently.

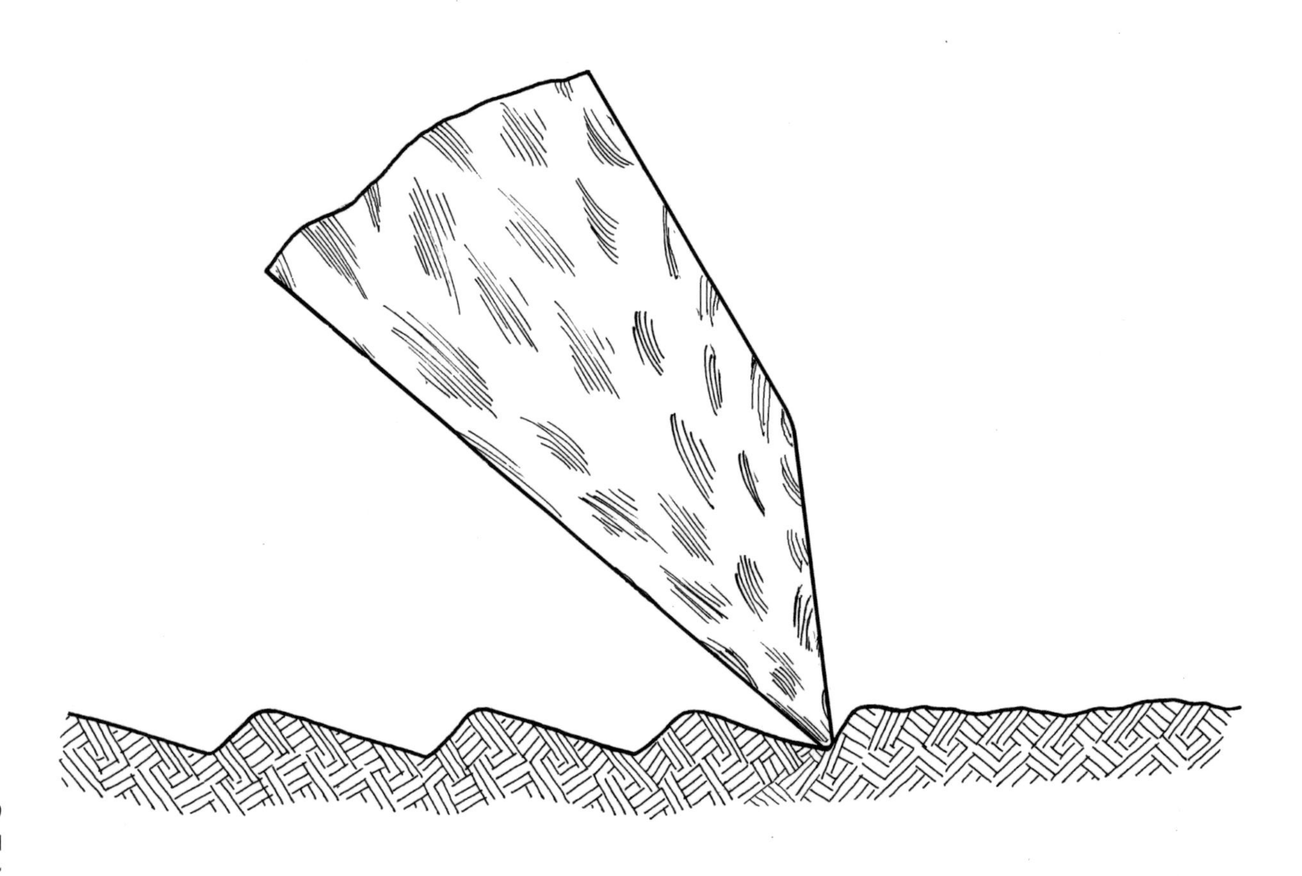

Fig. 20 Cross-section of a chisel track.

My hammer-hardened copper chisel works well on Giza limestone, but is quickly blunted by the Tura stone. It's hard to believe this type of chisel made the tracks in figure 19. I have made similar tracks, however, using my wooden mallet against a *flint* nodule with a flaked edge. Flint nodules occur naturally in limestone deposits, and are ubiquitous in the sands around the pyramids.

Fig. 21 Flint nodule from Giza desert.

Legendary British Egyptologist Sir William Matthew Flinders Petrie (1853-1942) believed that short-handled wooden adzes with flint cutting edges were used to dress the blocks in Khasekhemui's cenotaph (Dynasty 2) at Abydos. Petrie, mentor of Howard Carter (discoverer of Tutankhamun's tomb), is famous for bringing scientific methodology to Egyptology.

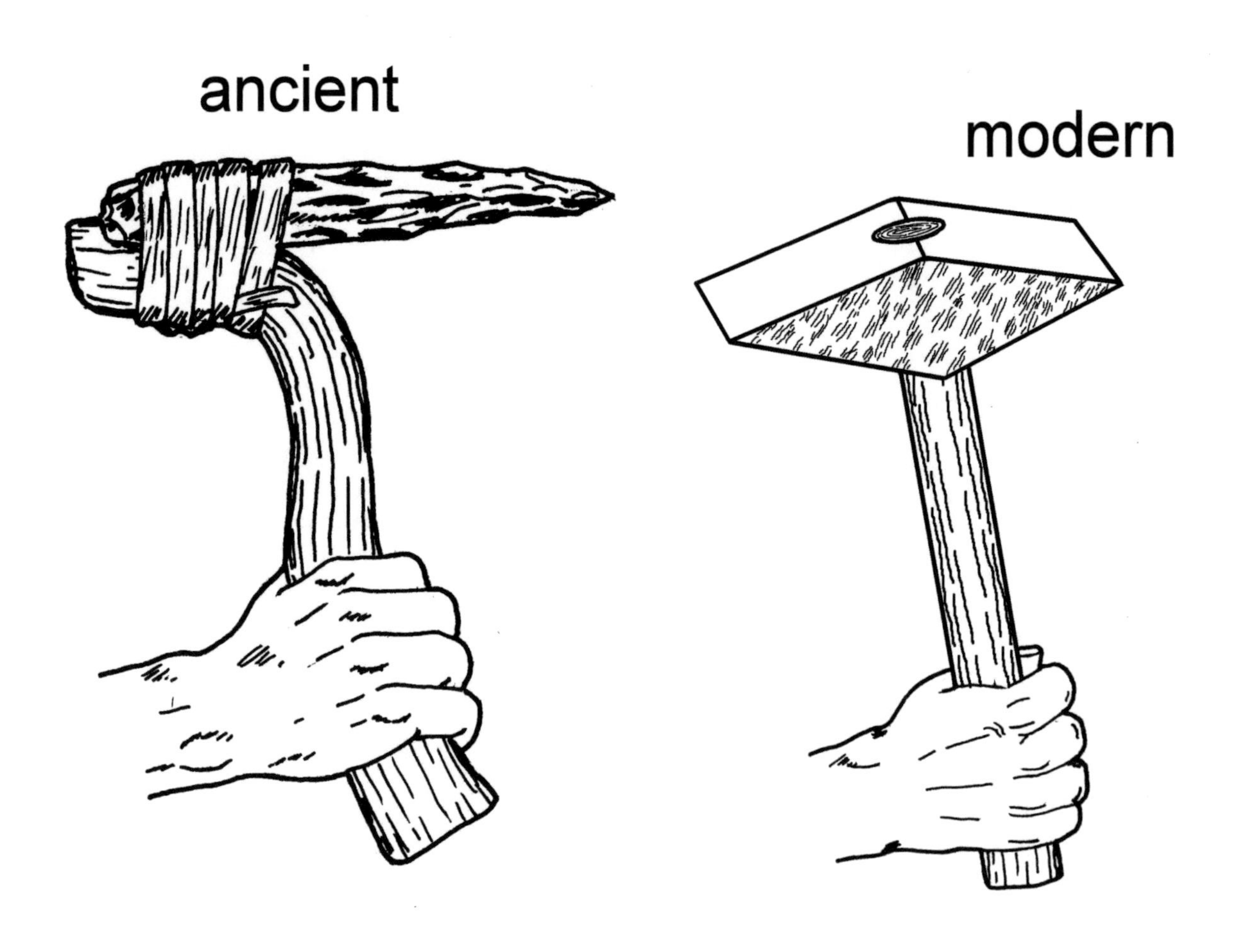

Fig. 22 The adze, ancient (flint) and modern (steel).

I have observed Tura quarrymen squaring blocks with a double-edged steel adze (fig. 22). Its cutting edges are about fifteen centimeters (6 in.) wide. Its wooden handle is roughly 46 cm (18 in.) long. Clarke & Engelbach wondered (1930, 17), as do I, if the ancient masons used something similar.

The flint-blade adze seems more practical for working all types of limestone than the copper chisel. The quarry face at the northwest angle of Khafre's pyramid at Giza is scored by curved grooves having radii of about a half meter (20 inches). These arcs would correspond nicely to adze-wielding human arms.

Fig. 23 Tool tracks on north face of quarry at NW corner of Khafre's pyramid.

Tracks on the pyramid stones could have been produced by the same technique. On the other hand, the marks on Khafre's northwest quarry (fig. 23) may be a result of the manner in which the face was progressively worked. A hypothetical depiction of this process can accommodate either chisels or adzes, as shown in figure 24.

Fig. 24 Working the face of Khafre's NW quarry.

Stone tools seem less sophisticated than metal ones, but are better suited, I think, for soft-stone cutting. Flint chisels are not only harder than copper, they are easier to shape and sharpen. I doubt these qualities would have been lost on the masons. Thus I believe both flint and copper tools were used to cut limestone, but which was preferred in a particular situation I cannot guess.

Granite

After local and Tura limestone, the third most-frequently encountered building material is the beautiful rose granite of Aswan. This stone was used more sparingly than limestone because it was much more difficult to obtain and work. It had to be shipped from the First Cataract, five hundred miles upriver. Aswan granite gets its pink color from large crystals of red microcline (potassium feldspar). It contains smaller amounts of clear quartz, biotite mica, and hornblende.

Fig. 25 Aswan rose granite.

Much of the granite taken from Aswan during the Old Kingdom may have been obtained from loose boulders. In the New Kingdom, a thousand years later, when mega-blocks were needed for obelisks and other colossi, quarrymen had to extract them from bedrock.

There is evidence that boulders were divided by wedge splitting. Ancient wedge slots can be seen at Aswan and on the rear edges of granite casing stones strewn around Menkaure's Pyramid at Giza.

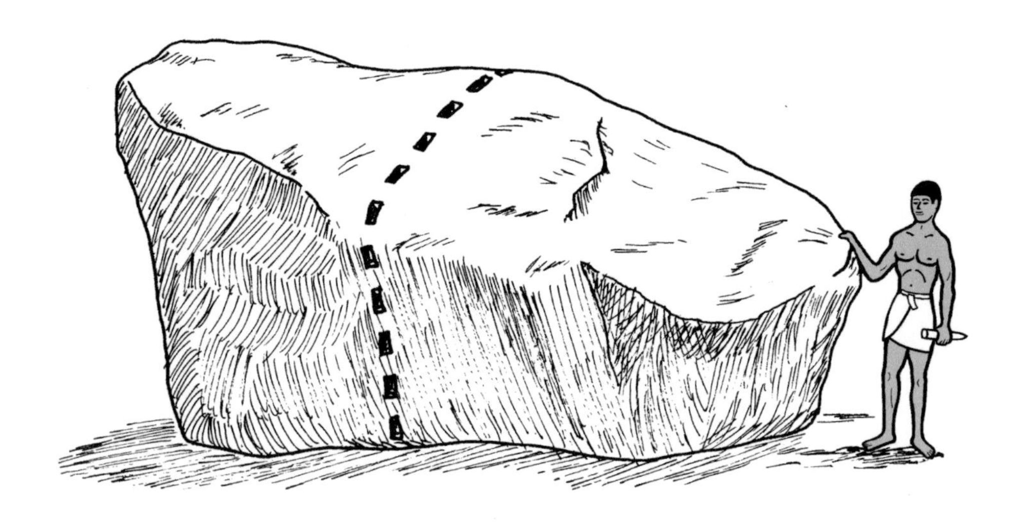

Fig. 26 Wedge slots in granite boulder.

The shape of the ancient wedge is unknown. Same for its material. and technique of use. Modern quarrymen place steel "feathers" on the sides of the wedge slot (or hole) to reduce frictional resistance to the entrance of the steel wedge (or plug).

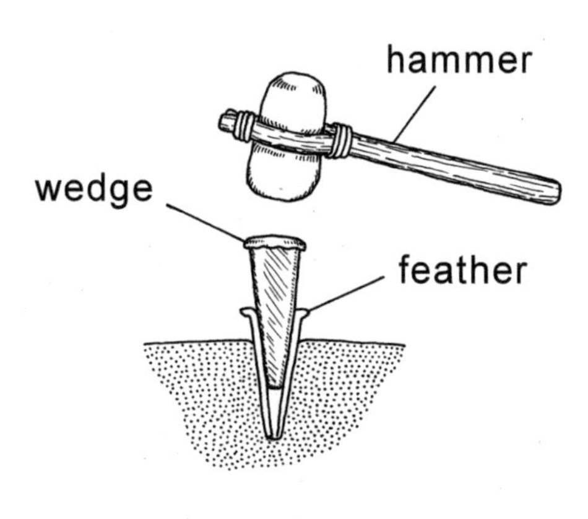

Fig. 27 Hypothetical ancient wedge-and-feathers combination.

Engelbach (1922, 4) examined wedge slots in the Aswan quarries. He discounted the idea that wooden wedges, expanded by the action of water, could have been used in these places because the sides of the slots were so smooth that wetted wooden wedges would "tend to spring out rather than exert a lateral force on the stone." He favored the idea that the wedge-and-feather method was used, "but the question is best left open for the present lack of conclusive evidence."

How the granite blocks were shaped is not so mysterious. To work granite the builders took advantage of a mineral property I did not mention before: ease of fracture. A hard mineral may also be brittle, and fracture upon impact. The large crystals of feldspar and quartz can be easily crushed by hammer-stones of a finer-grained rock of equal hardness.

At Giza the Egyptians shaped the granite casing blocks of Menkaure's pyramid with granodiorite hammer-stones about the size of your fist. In 1987 these tools could be found in the sands near the northeast corner of the pyramid.

Fig. 28 A granodiorite hammer-stone.

Using one of these on a large piece of granite I was able to make faster progress than I expected. Light blows at a steady pace worked best. Tap-tap-tap. The coarse crystals were crushed to a fine powder which I could sweep away with my free hand. The hammer-stone was virtually unaffected.

To extract the famous Unfinished Obelisk at Aswan (New Kingdom) quarrymen used large balls of compact dolerite, grasped by both hands, to pound out surrounding trenches in the rock (Engelbach, 1922).

Fig. 29 The Unfinished Obelisk of Aswan. Photo by Sheri Moores.

Stone hammers and wedges were not the only tools used on granite. The Egyptians could *saw* and *drill* granite and other hard stones like basalt and quartzite. Lucas believed these operations were performed with reciprocating saws and tubular core drills of copper, both being charged with quartz sand as cutting media. My experiments using this technique were exasperating. The edge of my bow drill was eroded much faster than the granite was pierced.

Fig. 30 Core-drilled door pivot, Khafre's Valley Temple. Hole diameter about 7cm.

Fig. 31 Tubular bow drill.

A Dynasty 5 scene in Sahure's temple at Abusir shows a workman using a weighted drill bit to hollow out the interior of a stone vase. He is spinning the drill-shaft by hand. Unfortunately, we are in the dark as to the cutting edge of the bit, and if an abrasive medium was employed.

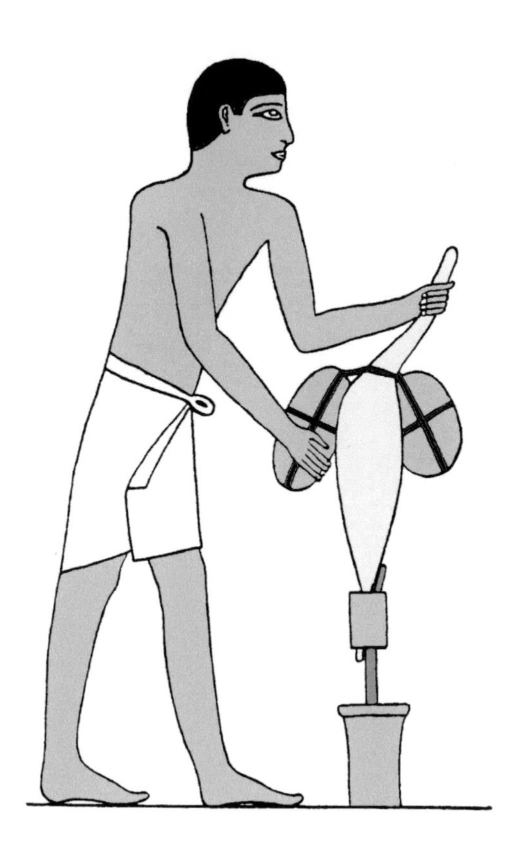

Fig. 32 Vase drilling scene from Dynasty 5 temple of Sahure at Abusir. Redrawn from Clarke & Engelbach (1930).

Petrie proposed that the copper saws and tube drills had corundum or diamond stones set into the metal, thereby forming jeweled cutting edges. His reasoning is based on many examples of cored holes and sawn faces that contain deep scratches. In *Tools and Weapons* (1917, 44), he describes a red granite core found at Giza:

> On this a continuous groove of the drilling point can be traced for several rotations, forming a true screw thread, and showing a rapid descent of the drill. The grooves run continuously across the quartz and felspar [feldspar] crystals without the least check; as the felspar is worn down (by rubbing) more than the quartz, the latter crystals stand highest; yet the grooves run with an even bottom through a greater depth of quartz than of felspar. Every mechanician who has examined this agrees that nothing but a fixed point could have cut such grooves.

Lucas believed these scratches were a result of particles of a loose grinding medium, such as emery or quartz, becoming embedded in the soft copper tube or blade.

Fig. 33 Horizontal grooves on north end of red granite coffer in King's Chamber of Khufu's pyramid.

Basalt

The pavement of Khufu's mortuary temple on the east side of his pyramid is black basalt. Robert Hamilton, Senior Research Physicist at the Manville Service Corporation in Denver, Colorado, analyzed two sample chips I sent him in 1989. He described them as follows:

> Pyroxene diabase – medium grained, hypidiomorphic-granular, sub ophitic to intergranular. 65% plagioclase feldspar (labradorite in composition), 30% augite, 5% ilmenite. Five to thirty percent of the intergranular material originally consisted of glass which has now been altered to a mixture of clays and iron oxides. The browner of the two samples contains more altered glass than the blacker one. Despite their difference in color the two samples came from the same rock unit.

According to James Harrell and Thomas Bown, writing in the *Journal of the American Research Center in Egypt* (*JARCE*, 1995), the source of this material is the quarry at Widan el-Faras. From there the blocks were hauled over a flagstone-paved road "to the shore of ancient Lake Moeris . . . where they were loaded onto barges at a quay. The barges sailed across the lake and through the Lahun-Hawara gap to the Nile River."

Extensive sawing evidence can be seen on these paving stones. See figures 34 and 35. The marks are perplexing as to how and why they were made.

Fig. 34 Saw marks on basalt paving stone of Khufu's mortuary temple.

Fig. 35 Partly sawn basalt block, Khufu's mortuary temple.

I authored a paper in the 1991 issue of *JARCE* wherein I proposed that these marks were produced by a copper-blade drag saw that was fed quartz sand as a cutting medium. I favor the drag saw because the saw marks are straight. There are no concave curves as would be expected from a circular blade and no convex curves that would be expected from a wire saw. Plunge cuts on some blocks indicate the tip of the blade was only about three millimeters wide! The sides of the plunge cuts are smooth – no scratches.

The sawing method remains a mystery.

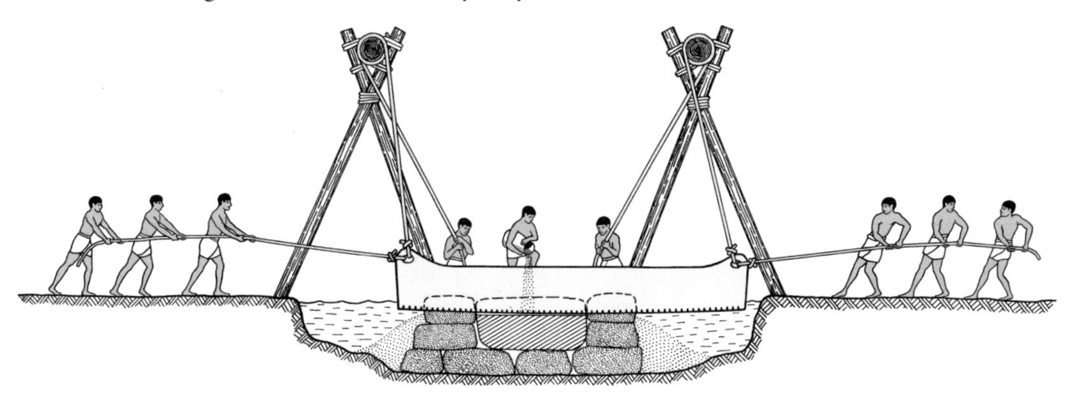

Fig. 36 My hypothetical drag-saw setup.

Iron tools?

On the use of iron tools in the pyramid age, Lucas said (1962, 236):

> The fact that a few specimens of iron of early date have been found has been used to support this argument [that the pyramid builders used iron tools], and it is stated that it is only on account of the easily oxidizable nature of iron that tools and other objects of this metal have not been discovered more frequently. Iron, however, although it does oxidize readily in damp soil, particularly if salt is present, is quite stable under the ordinary conditions that prevail in rock-cut and other tombs in Egypt into which water has not penetrated, and the fact that some few specimens of iron have survived is proof that had there been other examples under similar conditions these too would have lasted. It should not be forgotten also that iron when it oxidizes does not disappear, but is converted into a compound that it is not only permanent but which, on account of its

reddish colour and of its greater volume than that of the original metal, should not escape observation.

One of the pieces of iron Lucas is talking about is a flat rectangular plate found on Khufu's pyramid in May 1837 by Mr. J.R. Hill. He found it while clearing blocks from the exterior terminus of the southern air-duct that originates in the King's Chamber of the pyramid. Working for retired British army officer Col. R.W. Howard Vyse (1784-1753), Hill asserted in an affidavit (Vyse, 1840, 275) that the piece of iron was found in an "inner joint" under two layers of core masonry. He said "that no joint or opening of any sort was connected with the above-mentioned joint, by which iron could have been placed in it after the original building of the pyramid."

Petrie, upon examining this item in the British Museum, said (1885, 85) that it "has a cast of a nummulite on the rust of it, proving it to have been buried for ages beside a block of nummulitic limestone, and therefore to be certainly ancient."

Lucas remarked (1962, 237):

"Although the statements of the finder (Mr. J.R. Hill) and others, who examined the spot at the time, are very definite and precise and not lightly to be disregarded, it seems more probable, since the iron has been proved not to be meteoric, that it is of recent date and that it had been lost down a crack in the stone facing of the pyramid when this was being removed for use as building material in modern times, long before Vyse's work."

Note that "recent date" and "modern times" referred to by Lucas could be as early as the Middle Ages, when most casing stones were stripped from the pyramids. Nine hundred years would be enough time for the cast of a nummulite to have formed on the iron surface, especially if water from (admittedly infrequent) rainfalls could reach it. Further, the shape of the iron sheet suggests, as Lucas inferred, that it could have served as one of two feathers used in wedging casing stones away from the pyramid.

With no other evidence of iron tools during the pyramid age we must conclude that they were not used. Nor have any bronze tools from that period been recovered.

Mortar

The builders put gypsum mortar between the largest blocks of the most solidly-built pyramids, particularly those at Giza. Gypsum is calcium sulfate dihydrate. They employed two types of gypsum mortar, each for its own purpose. To facilitate placement of the large casing stones they used an almost pure (and white) variety (that today we call "plaster of Paris") as a lubricant while sliding the blocks to their final positions.

A more coarse gypsum mortar, containing crushed pottery, sand, and limestone, was used to fill spaces between backing and packing stones. This mortar often has a pinkish color due to the crushed pottery it contains. In this application it served a stabilizing function (preventing movement) more than a cementing one (gluing blocks together). Figure 37 shows the mortar between two backing stones on the east side of Khufu's pyramid. Note the tool marks on the block at left.

Fig. 37 Coarse mortar between backing stones on Khufu's pyramid.

Fig. 38 Coarse mortar used in Khufu's pyramid.

My samples of mortar from Khufu's pyramid are light and porous, perhaps because they have been exposed to rainwater for several thousand years. Dr. Iskander, in analyzing relatively protected samples in Khufu's boat pit, reported in *The Cheops Boats* (1960) that the coarse mortar around the limestone filler slabs between the blocks and pit walls was composed of roughly eighty percent gypsum, the rest being filler minerals. He found that the mortar between abutting ceiling blocks was nearly 99 percent gypsum.

Properties of Egyptian building stones

I collected a number of stone samples to measure density of the four types of building stones mentioned above, so I will conclude this chapter with a report of those findings.

Table 2 – Specific gravity and density of Egyptian building stones

Material	Specific Gravity	Density, lbs/ft3	Density, kg/m3
Fine white limestone	2.055*	128	2055
Local Giza limestone	2.313**	144	2313
Rose granite, Aswan	2.664	166	2664
Black basalt, Giza	2.851	178	2851

* Average of thirteen samples that ranged 1.970 – 2.168
** Average of eleven samples that ranged 2.235 – 2.362

For these measurements I used an analytical balance with a below-balance hook that allowed weighing each sample in air and distilled water. The difference in weight is the weight of water displaced (by Archimedes' principle). Knowing the weight of water displaced determines the volume of the sample because pure water weighs one gram per cubic cm.

I will refer to these densities when I talk about block weights throughout this book, but please understand that these values are imprecise for two reasons. The first is that my sample chunks were small (50-200 gm). The second relates to limestone in particular. When using the method of displaced water to determine volume, as soon as limestone is submerged it rapidly begins to absorb water. The effect is that the sample becomes heavier with each passing second. Thus the measurements had to be taken quickly, within a few seconds. My reported densities of the limestone samples are then, if anything, too high.

3

TRANSPORTING THE STONES

Land Transport

The builders transported blocks over smoothed roads from quarries and quays. The roads do not have large gradients. The steepest I know of is the "approach road" (see fig. 76) that runs nearly parallel to the causeway leading to the Meidum Pyramid. According to Petrie, it inclines slightly more than six degrees. The upper portion of the causeway to Khufu's pyramid is still intact. Its slope is about four degrees. The modern road from the Sphinx to Khufu's pyramid, possibly an ancient supply route, rises at an angle slightly greater than five degrees. In papyrus Anastasi I, the scribe Hori asks for an estimate of the number of bricks required to construct a ramp 730 cubits long and 60 cubits high (Engelbach 1922, 35). Such a ramp would slope at 4.7 degrees.

At Sneferu's Northern Stone Pyramid at Dahshur (the Red Pyramid) two transport roads approach the pyramid from quarries to the southwest. Both roads fade out some two hundred meters from the pyramid, but if extended would converge on a point about seventy-five meters south of its southwest angle. That could be a construction clue, but I don't know what it means.

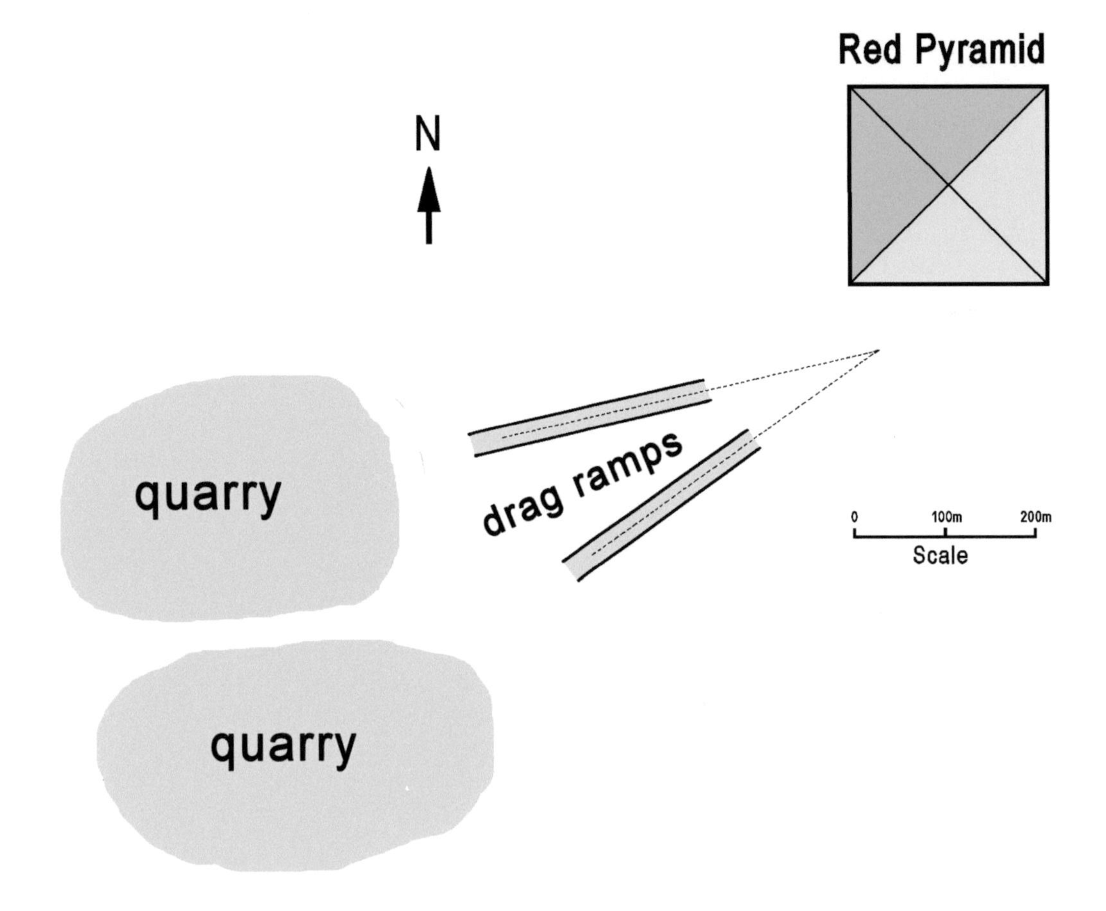

Fig. 39 Transport roads at the Northern Stone Pyramid of Dahshur. Adapted from Klemm (2010, 61).

The builders employed several methods for transporting materials on land. The two most widely used were probably carrying and dragging on sleds.

Carrying

In the article I mentioned in chapter two, Harrell and Bown described an Old Kingdom road used to transport basalt blocks from the quarry at Widan el-Faras to Lake Moeris. The road is about seven feet wide. It is paved with limestone slabs about twenty inches across. The authors reported that "the paving stones show no obvious evidence of wear," i.e. no marks of anything having been dragged over them. They suggested two possibilities for the transport method: Either the blocks were carried by some means, or wooden planks were laid in front of block-bearing sleds as they were dragged along the road.

Carrying is the simplest explanation. If wooden planks had been used we would expect them to have been embedded in the road. Otherwise they would have been sliding askew as sleds were drawn over them. Keeping them properly positioned would have been difficult. Besides, planks would have provided little (if any) reduction in friction compared to bare paving stones. No, I believe the paved road provided firm support for the people who carried these blocks using wooden (or bamboo?) poles.

An obvious question follows: Was carrying the typical method used to transport pyramid stones to, and possibly up, the pyramids? It depends on the size of the block. Most of the stones in Dynasty 3 pyramids, and many in Dynasty 4, are man-portable. So were the mud-bricks in mastaba tombs of Dynasty 3 and before — sleds not needed. Mortar components, limestone filler chunks, and other sundry items were almost certainly carried via human legs.

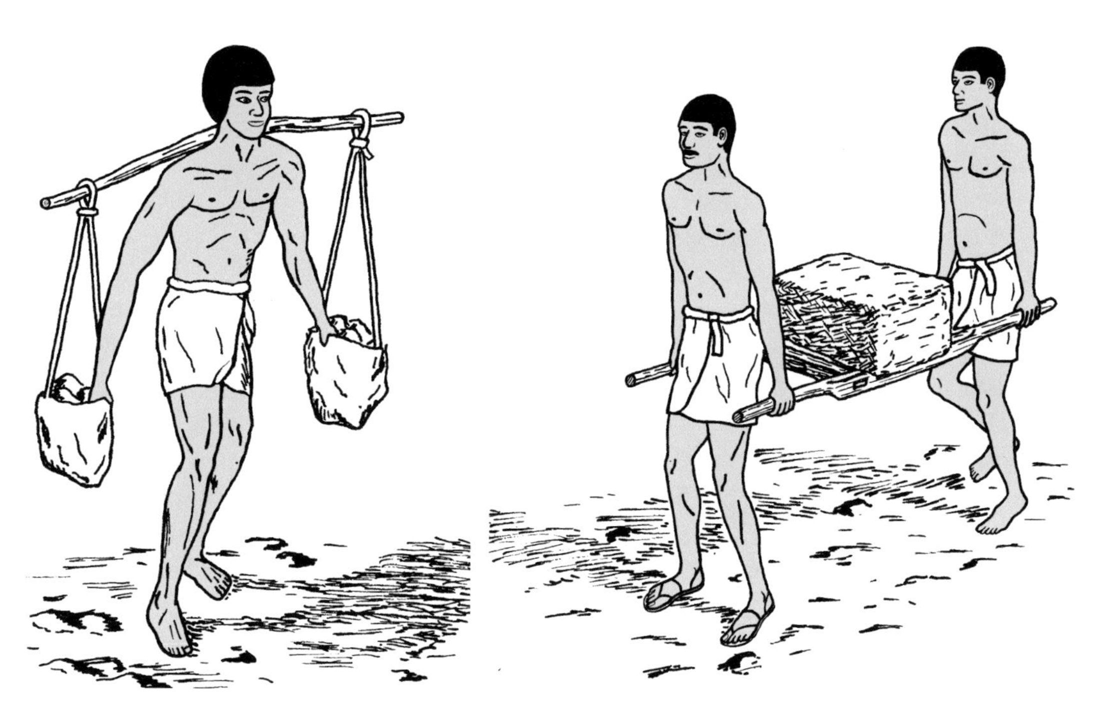

Fig. 40 One-person balanced-pole carry.

Fig. 41 Two-man stretcher carry.

Fig. 42 Eight-man pole-carry.

But block size increased with each new pyramid, at least through Dynasty 4. The blocks in Djoser's Pyramid range from about 160-500 kg (350-1100 pounds). In Khufu's pyramid, the average had risen to 2,350 kg, or 2.6 tons.

I figure the practical limit for carrying a stone block is about one ton (2,000 pounds). If limestone, it could look like a block with dimensions of 38 x 63 x 94 cm (1.5 x 2.5 x 3.7 feet). Allocating each workman a hundred pounds of the 2,200 pound load (2,000 pound block + 200 pound carry-lattice) twenty-two people would be required. A carry-lattice of crossed poles (like #) is needed because twenty-two people cannot fit close enough to the block to hold it directly. Since a carry-lattice is not weightless, and bending stresses on the poles increase with distance from the load, the efficiency of this method drops as the load increases. The lattice becomes an ever-larger part of the load, demanding additional effort to carry just the lattice. An ever-increasing space is needed for the carriers. A four ton block would need about 80 carriers in eight rows of ten, and occupy an area of about 10 x 12 meters (33 x 40 feet). Hard to picture that in practice.

Though impractical for the heaviest loads, carrying is more efficient than dragging because less energy is lost to friction and less preparation of the transport path is needed. A reasonable scenario is that blocks and chunks reaching perhaps 555 kg (1,220 pounds) were carried by teams of up to fourteen people.

Dragging on sleds

There is considerable evidence that heavy objects transported over land in ancient Egypt were dragged on wooden sleds. Two sleds have been recovered and preserved. Both date to the Middle Kingdom. The smaller, just under two meters long (about six feet), was unearthed close to the pyramid of Senusret I at Lisht. The larger is four meters long and one meter wide. It had been buried in a brick vault near the pyramid of Senusret III at Dahshur.

Fig. 43 Senusret I sled (Dynasty 12) from Lisht. From Hayes (1953).

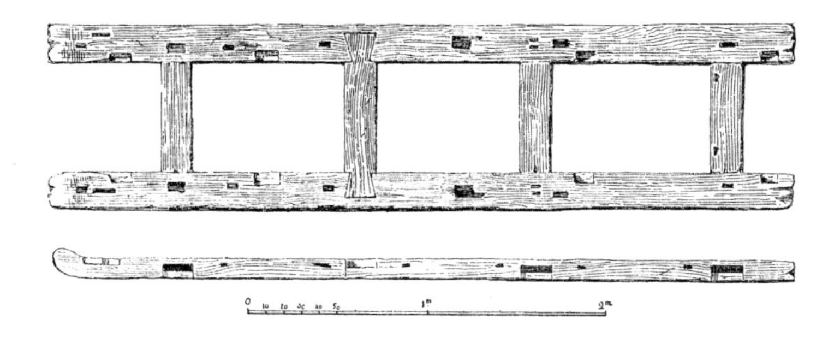

Fig. 44 Senusret III sled (Dynasty 12) from Dahshur. After De Morgan (1895).

The Lisht sled seems a more standard design because it appears in tomb paintings through at least the reign of Ramesses II (temple at Abydos), seven hundred years later. This type is characterized by upturned runners at front, similar to snow skis, and chamfers (beveled corners) at the rear undersides. The larger sled lacked rear chamfers, as does a variation depicted in a Dynasty 18 rock carving at Tura. Here a sled is drawn by oxen, motive power Egyptologists believe was not employed in the pyramid age. I'll propose a purpose for the upturned runners and chamfered corners in chapter nine. Another example of the standard sled can be seen in the tomb of Ti, Director of the Hairdressers of the Great Palace during the fifth dynasty. His tomb is at Saqqara.

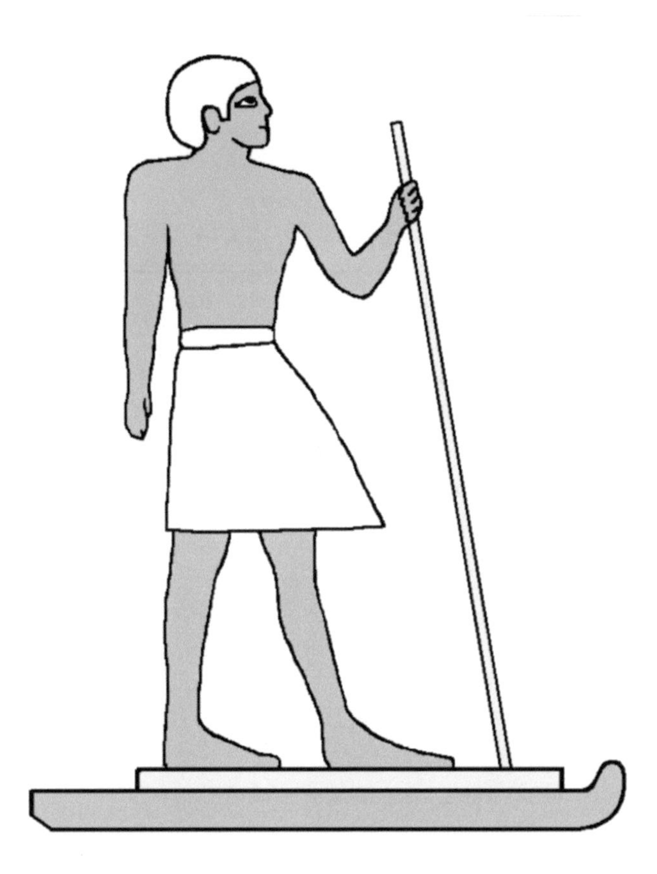

Fig. 45 Statue of Ti mounted on sled, 5th dynasty, Saqqara.

The earliest sled evidence comes also from Dynasty 5. Carved in relief on the wall of the causeway of Unas' pyramid is a barge carrying two stone columns. The columns are strapped to sleds that lack upturned runners. The sleds sit on transverse beams that probably served both to distribute their loads and as spacers between which maneuvering levers could be inserted.

Fig. 46 Transport of columns for temple of Unas. Adapted from Lauer (1976, plate 125).

Important stone-conveying information came to us via a wall painting in the tomb of Djehutihotep at el-Bersheh (Newberry, 1893). Djehutihotep governed the Hare nome during the Dynasty 12 reign of Senusret III, about 1878-1840 BCE. Fortunately, the famous scene was copied by Egyptologists before it was mutilated by vandals in 1891.

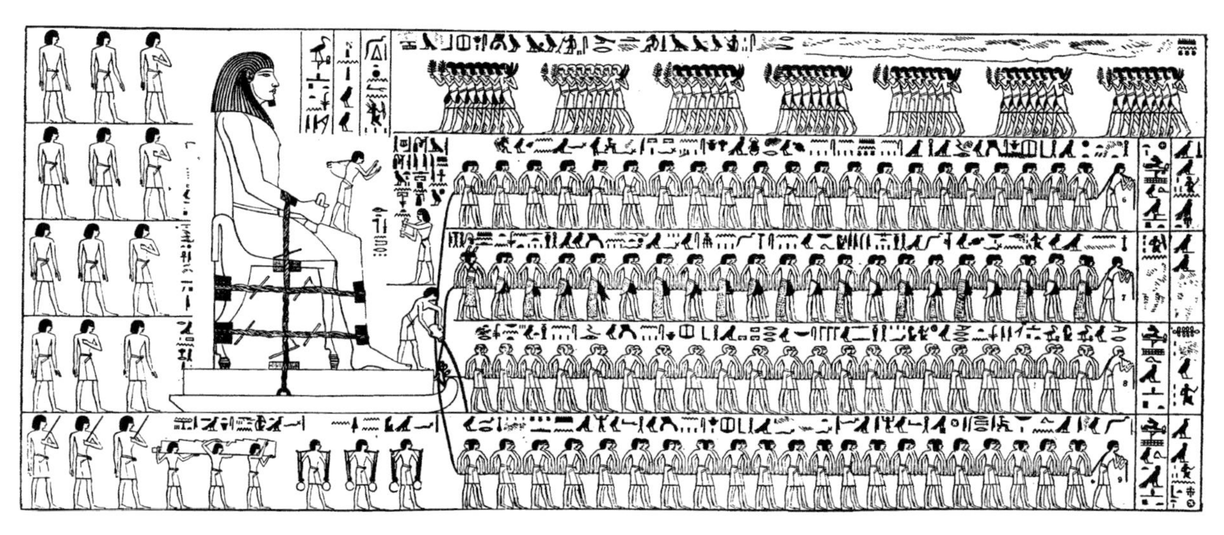

Fig. 47 Transporting the statue of Djehutihotep. From Newberry (1893).

The painting describes the joyous arrival of the statue of the governor. The monolith, we are told, is "stone of Hatnub" (calcite), thirteen cubits (22 feet) high. These data would yield a weight of about 85 tons as pictured. Inscriptions also say "Behold, this statue, being a squared block on coming forth from the great mountain, was more valuable than anything." The rough block as extracted from the quarry at Hatnub, 26 kilometers (16 miles) southeast of el-Bersheh, then probably weighed close to 100 tons.

The statue is lashed to a wooden sled drawn by 172 men in four rows. The top row is described as "The youths of the west of the Hare nome…" The second row is comprised of "The youths of the warriors of the Hare nome…" In the third row we find "The courses of the priests of the Hare nome…" The fourth row is described as "The youths of the east of the Hare nome…"

Above the rope-haulers are supporters and the inscription "The Hare nome is in festivity, its heart is glad; its old men are children, its youths are refreshed, its children jubilate; their hearts are in festivity when they see their lord, the son of their lord as a favor of the king, making his monument."

The man standing on the knee of the statue is described as "beating time for the troops by the signal-giver of Djehutihotep, beloved of the king." A man nearby seems to be smacking two objects together, perhaps amplifying the clapping of the signal-giver. Under the statue we find "workmen carrying water."

The "troops" is a translation of the word *zamu*, which Newberry interprets as young men trained for united labor. In this discipline the Egyptians must have been especially proficient, perhaps unequaled. The "time beater" shows that the Egyptians knew that greater force can be applied by a group of people surging forward in unison, the old "heave-ho."

A man standing on the base of the statue pours water on the ground in front of the sled. Water would greatly reduce the required pulling force if the road had a clay surface. Otherwise the transport scene would be inaccurate. At least three times as many men would be needed to drag the statue along a dry, level road. If sled runners bear on stone, wooden sleepers, or deep sand, water won't reduce the dragging force. This requires a short digression to *friction*, the force that resists the sliding of one object against another.

Suppose it takes a force of thirty-five pounds to drag a hundred pound object along a smooth, level surface. The *friction coefficient* is a number, or ratio, that compares the pulling force to the weight of the object being pulled. In this example a 35 pound force divided by a 100 pound weight equals 0.35, the friction coefficient.

The friction coefficient is pretty much constant for one material sliding against another, regardless of the area of the faces in contact and the weight of the sliding object. If both surfaces have significant bumps that interfere with each other (interlock), *sliding* friction would hardly apply. Table 3 shows friction coefficients for material combinations I investigated.

Table 3 – Friction Coefficients

Sliding materials	Angle of Friction (degrees)	Coefficient of Friction
Tura limestone on Giza limestone	26	0.488
Same but water wet	29	0.554
Aswan granite on Giza limestone	25	0.466
Same but wet clay between faces	5	0.087
Tura on Tura, Tura powder between faces	37	0.754
Oak on oak, cross grain, static	19	0.344
Oak on oak, cross grain, sliding	16.5	0.296
Oak on limestone	25	0.466

I found these values using the setup shown in figure 48.

Gradually increase the angle of the base material (angle θ) until a light push on the block starts it in motion at constant speed. This angle, at which motion impends, is called the *angle of friction*. By fortunate coincidence, the tangent of the angle of friction equals the friction coefficient.

The static (starting from rest) friction coefficient is slightly higher than that of sliding friction. The effect is that once an object is moving, the friction force decreases.

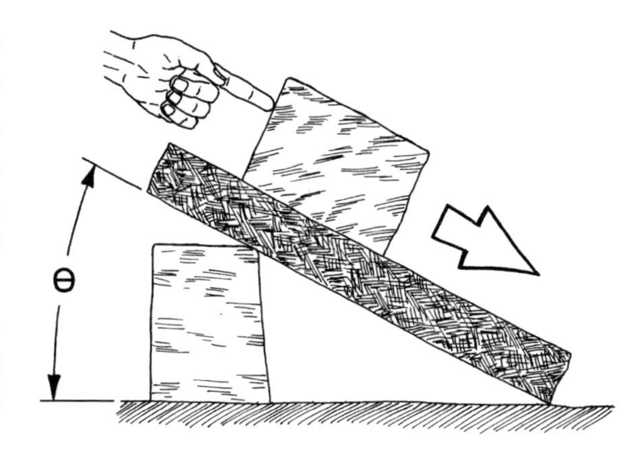

Fig. 48 Friction test apparatus.

My point: If we can determine the nature of the materials in contact, the weight of the object being dragged, the gradient along which it is being dragged, and if a lubricant is employed, we can calculate the force needed to move the object.

Back to Djehutihotep.

Newberry wondered if pouring water ahead of the sled was a ceremonial act, not necessarily for reducing friction. His idea could be valid. The same procedure can be noticed in other paintings where much smaller objects are drawn by only two or three men. If lubrication was intended it would seem less troublesome to forget the water and add a few more men on the rope. In the case of Djehutihotep, though, a water-lubricated clay-surfaced road is consistent with the elements depicted.

The beam carried by Djehutihotep's workmen may have been a sleeper, but why its upper surface is serrated I cannot guess.

Evidence for use of wooden sleepers was discovered during the excavation of the Middle Kingdom pyramids of Senusret II at Lahun and Amenemhat I at Lisht.

Fig. 49 Wooden sleepers in quarry road at Lahun pyramid of Senusret II. Drawn by author from photo in Lahun II, Petrie, et al (1923).

A ramp was found beside a brick wall bounding the pyramid temple of Amenemhat I. The ramp was about 6.5 meters wide, with brick retaining walls 90 cm thick. Depressions in the surface of the ramp suggest that sleepers had been embedded.

Fig. 50 Depressions for sleepers in ramp at the pyramid of Amenemhat I. Drawn by author from Mace (1914).

Dieter Arnold, in *Building in Egypt* (1991), said of another Middle Kingdom pyramid, also at Lisht:

> The excavations of the pyramid of Senwosret I at Lisht, conducted by the Metropolitan Museum of Art, New York, revealed numerous remains of transport roads approaching and surrounding the pyramid from all sides. The more or less horizontal transport roads consisted of 5-meter-wide tracks of limestone chip and mortar into which rows of beams (reused boat timber) were inserted in such a manner that their upper face disappeared under the road surface. Since the reused timber was too short to cover the road, frequently two pieces had to

be positioned together with their ends overlapping in the center. The core and surface of the roads had a cement-like hardness.

Dragging sleds over sleepers laid at right angles to the direction of transport is more economical and expedient than making clay-surfaced roads. The latter would have been needed only for the heaviest blocks. Where sleds were drawn over smooth rock surfaces no sleepers were required.

Another point about sled dragging concerns the position of the hauling ropes with respect to the haulers. I am amused when I see TV programs about the pyramid builders where block haulers have the dragging ropes slung over their shoulders. Have you ever participated in a tug-of-war? Where did you hold the rope? Chances are you had the rope at your waist and you were leaning backwards as you pulled. Why? Because you can generate much more force with that arrangement.

What is the easiest way to haul a sled up an incline? I found a section of my brother's asphalt driveway that slopes at 12 degrees. I discovered that the least fatiguing method developing the highest hauling force was to attach the rope to my waist and walk *backward* up the slope. By this means I was able to generate a force of seventy pounds rather comfortably. Applying this method to a hauling crew, perhaps each member would attach himself to one of the main hauling ropes with a small rope or sling that could be attached or detached as needed. The *prusik* knot used by mountain climbers is one that works this way.

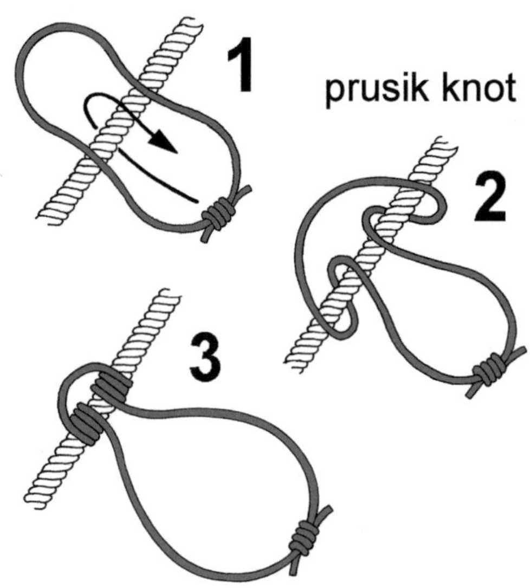

Fig. 51 The prusik knot.

Dragging on the ground

Could blocks have been hauled directly across the ground? In his investigation of Inca stonework at Ollantaytambo, Professor Jean-Pierre Protzen found evidence (1993, 176-182) that the megalithic blocks of the fortress had been dragged from quarries in the hills across the Urubamba River by exactly that method. Seasonally the river is only a meter deep, so it was not much of a hindrance. The blocks had "more or less parallel striations" on just one of their two broadest faces. Protzen imagined that "a big net, into which four or more long ropes were woven, was thrown over the top and sides of the block." By this means, Protzen thought the largest blocks (c. 117 tons) were drawn across the ground by up to 1,800 workmen.

I doubt the Egyptians used this method to transport blocks to the pyramids. For one thing, evidence for sleds is extensive. For another, if hardness of the path varies, the stones would tend to dig in on the softer portions. That said, on earthen ramps, as I'll cover in chapter ten, blocks could have been hauled directly over embedded stone slabs. That would have obviated logistical problems related to sled use.

Rollers and wheels

Might the Egyptians have used rollers or wheeled transport?

Rollers can greatly reduce friction. The reduction depends on two things, diameter of the roller and the materials of the roller and the surfaces against which it rolls. Rolling resistance decreases in direct proportion to increased roller diameter - larger rollers yield less resistance. The second factor is not as straightforward. The harder, smoother and more resistant to deformation the materials are, the greater the reduction in rolling friction. In other words, a roller is most effective if it remains round (not crushed by its load) and the surfaces it touches remain flat. Also, the crushing force on a roller is reduced by using more than one roller, the more the better.

Petrie unearthed two wooden "rollers" during his excavation of Senusret II's pyramid at Lahun. I doubt these were used for transport because they are too small (5.5 cm diameter by 23 and 33 cm long) to have been easily positioned for large objects.

Fig. 52 Lahun rollers. Drawn by author from photo in Tools and Weapons, Petrie (1917).

Arnold suggested (1991, 116) that rollers were employed to maneuver blocks close to their final places in *wall* construction, but presents no evidence for this conjecture.

Recently I watched a program on the Science Channel where researchers tested the idea of roller use. On *level* ground they placed wooden beams like rails of a train track. They put rollers between the sled and the rails. The rollers appeared to be *metal* pipes about ten centimeters (4 inches) in diameter. The video made it look like this procedure worked, but we were shown only a few seconds of sled movement, and nothing of the time it took to reposition the rollers. Beware of selective editing supporting a particular view!

If the builders understood the mechanics of rollers they must have considered roller use impractical or unnecessary. Because rollers do not appear in any transport scene it is less presumptuous if we doubt their use.

Of wheel evidence we have a unique painting from the Dynasty 5 mastaba of Kaemhesit at Saqqara. The scene shows men armed with maces climbing a ladder having small wheels at it base. It's the earliest and only representation of wheels (or a ladder) until we reach the New Kingdom. Lacking any other evidence we must make the same judgment for wheeled vehicles that we made for rollers: they were not used for transport in the pyramid age.

Transport over water

In the spring of 1954 a group of workmen supervised by Egyptian archaeologist Kamal el-Mallakh (1918-1987) were removing a large mound of sand and debris that had accumulated against the base of Khufu's pyramid. They were working on the south side, the last to be cleared. As they uncovered the boundary wall that runs parallel to the base of the pyramid they found something unexpected. The wall rested not upon bedrock of the plateau but on a layer of compressed earth, mud and rubble. To Mallakh's mind this "dakkah" had no purpose unless it was to conceal something.

Quickly digging through the compacted material, the workmen began to expose a row of forty-one limestone beams sealed with gypsum mortar. They found s similar group beginning a few meters to the west.

Mallakh's workmen, given permission by the Antiquities Service, completed a hole through block twenty-two of the eastern group on 26 May. Aided by sunlight reflected by his shaving mirror, Mallakh peered into the darkness.

The limestone monoliths formed the ceiling of a rectangular pit 31.2 meters (102 feet) long, 2.6 meters (8.5 feet) wide, and 3.5 meters (11.5 feet) deep. In the pit were the disassembled parts of a wooden ship, one of the greatest archaeological finds in history.

Because the pit had been well-sealed the boat had suffered only slight deterioration in the forty-five centuries of its interment.

Fig. 53 Khufu's funerary boat. Photo by Sheri Moores.

Khufu's boat, by far the oldest ever found, was painstakingly reconstructed by conservator Ahmed Youssef Moustafa over a period of fourteen years. It is housed in a museum built above the pit. From prow to stern she measures 43.5 meters (143 feet). She contains 1,224 wooden parts and assemblies, the majority identified as cedar of Lebanon. Many of the planks had symbols which indicated where two planks should abut. These may have been assembly cues for those who would reassemble the boat in Khufu's afterlife.

Because the craft was pointed west, some believe its purpose was to allow the pharaoh's spirit to accompany the solar god *Re* in his journey across the daytime sky. Others thought that the boat was simply Khufu's funerary transport. A project to disinter and preserve Khufu's second boat (possibly meant for the nighttime journey) was begun in 2009. This project is much more difficult because the western pit had not been sealed as well as the eastern pit, and the pieces are substantially rotted from the action of water, oxygen, and bacteria.

The design features of Egyptian ships have been compiled from tomb paintings, model ships, and a few full-size boats, the latter much smaller than Khufu's. Hulls were built with strakes abutting at their sides (carvel-built). There were no projecting keels. Larger craft, like Khufu's, required hull-strengthening ribs, but these were inserted *after* the hull was formed in what is called "shell-first construction."

Fig. 54 Symbols on abutting planks of Khufu's boat, in this example, the goddess Maat. Photo by Sheri Moores.

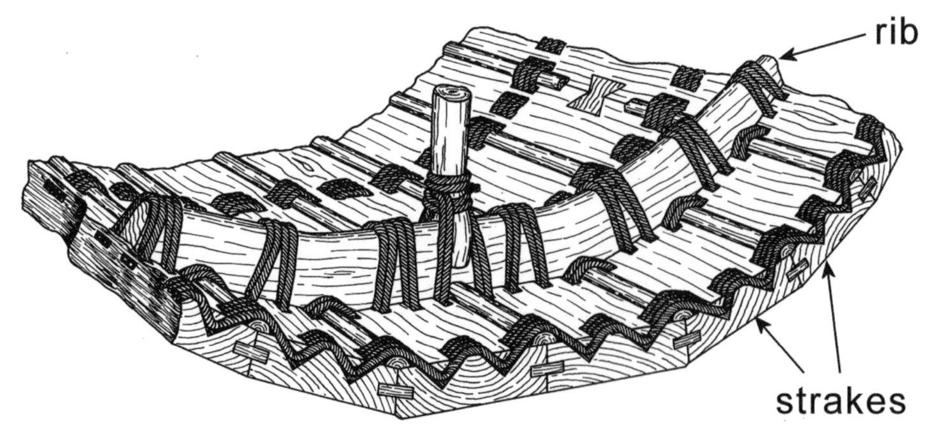

Fig. 55 Hull construction of Khufu's funerary boat.

The vessels were tied together with a fantastic interweaving of ropes that ran through V-shaped mortise cuts in the strakes and around the ribs. According to Paul Lipke (1984), Khufu's boat is the earliest example of a *sewn* ship, one with strakes lashed side-to-side and fore-and-aft. An advantage of this system, says Lipke, is it allowed disassembly of the ships for easier transport across the eastern desert to the Gulf of Suez and Red Sea.

The upswept bow and stern connected and supported the gunwales much like cables of a suspension bridge. Transverse deck beams kept the sides from bulging and provided cargo support. Vessels for heaviest transport had a *hogging brace*, a rope truss that resisted forces trying to break the ship in half when a wave temporarily supported it amidships. A hogging brace can be seen in the painting of Queen Hatshepsut's obelisk barge in her tomb at Dier el-Bahri.

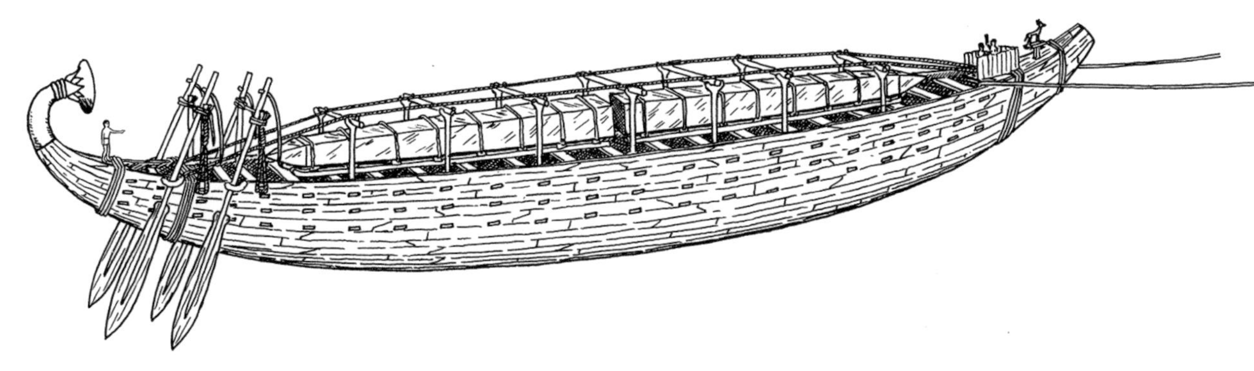

Fig. 56 Queen Hatshepsut's obelisk barge. From Naville (1908).

Khufu's boat was not intended for heavy transport, so its longitudinal brace is comprised of a central spine and two side girders (Lipke) at deck level. The boat had seen little, if any, time on the water. Nonetheless, it has been judged Nile-worthy and represents, experts believe, sophistication that must have evolved over centuries.

Fig. 57 The Egyptian cargo ship.

With few exceptions cargo ships were equipped with sail and oarsmen. Prevailing winds along the Nile are from the north, easing the trip upriver. When wind-power was lacking, row-power was not.

An important clue to understanding the technology of this period is found in the rigging of sailing vessels. Square sails were hung from horizontal yards. The yards were raised and lowered by halyards that ran through grid-like frames, perhaps of copper, attached to the tops of A-frame masts.

Fig. 58 Masthead detail of Egyptian sailing vessel.

Clarke & Engelbach (1930, 44) noted that if the Egyptians had known of the sheaved pulley they would surely have used it here. Nor do we see pulleys depicted in tomb paintings and relief carvings, or referred to elsewhere. From this we can conclude that the Egyptians were unaware of the friction reducing properties of the sheaved pulley.

4

SAQQARA

Saqqara lies twelve miles south of Giza, just west of the ancient capital of Memphis. Here a remarkable funerary complex was constructed for Djoser, the second king of Dynasty 3, whose reign commenced about 2630 BCE.

The architecture of Djoser's compound was without precedent, and its scale so ambitious that one wonders how much of its configuration stood in the original plan. Whose plan? Imhotep. His name and titles were found on a pedestal of one of Djoser's statues: Chancellor of the King of Lower Egypt, First One Under the King, Administrator of the Great Mansion, Hereditary Noble, High Priest of Heliopolis, Chief Sculptor, and Chief Carpenter. Of all the master builders to come, none surpassed his work and none were more revered. His ultimate victory may be the thoroughness with which he concealed his own tomb. It has never been found.

Imhotep is considered the inventor of ashlar masonry, the laying of squared blocks in tight-fitted courses. That is noteworthy enough because of the improved durability of buildings so constructed. But more amazing is the artistry and execution we usually associate with classical Greece, still two thousand years in the future.

The Egyptians had developed stone-working techniques earlier. They made limestone slabs to block crypt entrances in pit tombs of Dynasty 1. Floor paving, ceiling beams, and relief-carved stelae are dated before Dynasty 3. But nothing remotely comparable to Imhotep's work at Saqqara had ever been seen. Royal tombs had been constructed, like most important buildings, with bricks of Nile mud reinforced with chopped straw. These tombs were mostly in the low rectangular *mastaba* form and, though somewhat variable in detail, are not aesthetically impressive.

The opposite is true of Djoser's complex. A large rectangular area is bounded by walls in the form of bastions and curtains twenty cubits (10.5 meters) high. The walls are faced, like most of the interior elements, with beautifully wrought blocks of white Tura limestone. Within the enclosure are chapels, porticos, and pavilions embellished with statuary and engaged, fluted columns.

THE STEP PYRAMID

In the center of the whole is a stone pyramid, the first in Egypt. It has a rectangular base, 140 meters (459 feet) west to east and 118 meters (387 feet) north to south. It rises in six tiers to a height of 60 meters (203 feet). Hence its popular name, The Step Pyramid.

Fig. 59 The Step Pyramid of King Djoser.

The builders started by excavating a pit, seven meters (23 feet) square in plan, some twenty-eight meters (92 feet) into the rock. At the bottom they assembled a chamber of red granite blocks, apparently to contain the king's mummy. I say "apparently" because no human remains contemporary with any pyramid have been found.

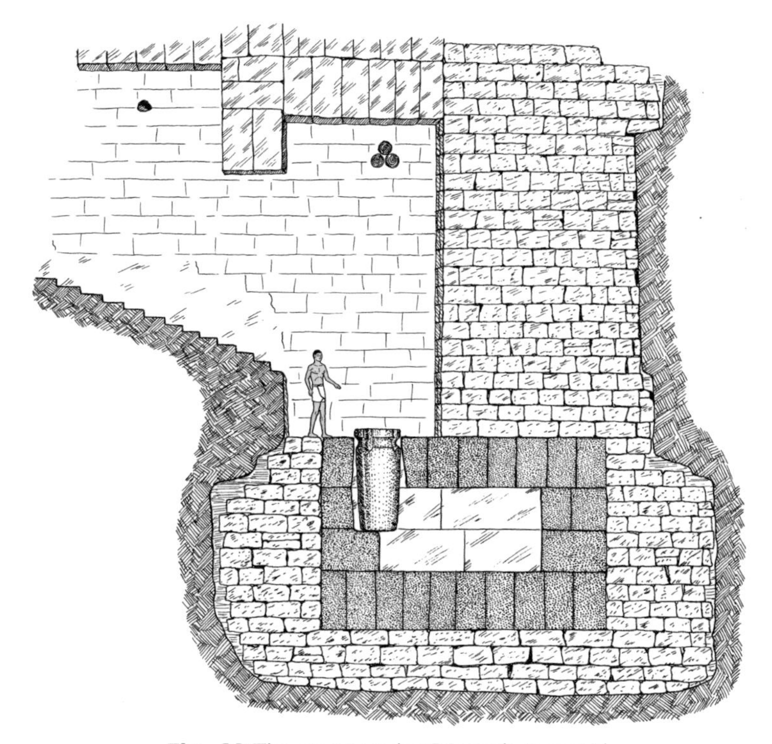

Fig. 60 The crypt under Djoser's pyramid.

The method the builders chose to seal the entrance to the vault was novel. The king's mummy would have been introduced through a hole in the ceiling. Into this hole a granite "cork" was lowered from a maneuvering room above. As reconstructed by French archaeologist Jean-Philippe Lauer (1902-2001), the maneuvering room had walls and ceiling beams of limestone.

The cork is a slightly tapered plug about a meter in diameter. Its length is almost two meters. It weighs about four-and-a-third tons. Notches and grooves near its top allowed purchase by lifting ropes. A hypothetical method of rope attachment is shown in figure 61. Noteworthy is that Dynasty 3 builders employed ropes which, in some configuration, could support a four-ton load.

The superstructure of the Step Pyramid evolved in five stages described by English Egyptologist I.E.S. Edwards (1909-1996) in *The Pyramids of Egypt* (1972, 42) The initial building (fig. 62-1) was a mastaba 63 meters (207 feet) square, of stone blocks laid in level courses to a height of about eight meters (26 feet).

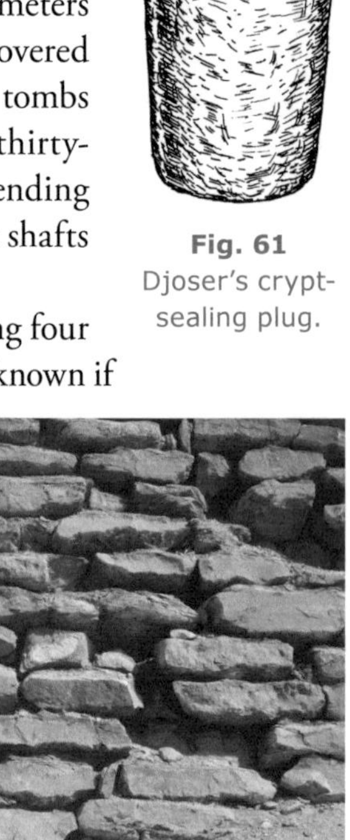

Fig. 61 Djoser's crypt-sealing plug.

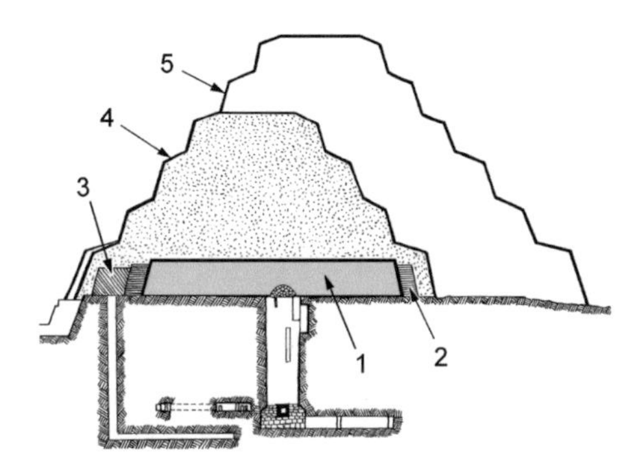

Fig. 62 The building stages of Djoser's Step Pyramid (looking south). From Edwards (1972, p 42).

The innovation must have made a good impression. Improvements ensued. A coating of masonry (fig. 62-2) about three-and-a-quarter meters thick (14 feet) was applied to each side of the mastaba. The blocks of this addition average about 0.07 – 0.15 cubic meters, and weighing roughly 160-340 kg (350-750 pounds). As you can see in figure 63, the blocks are not well-squared.

The east side was then extended about nine meters (30 feet). This addition (fig. 62-3) would have covered a north-south row of eleven pits which could be tombs of Djoser's relatives. The pits descend vertically thirty-three meters (108 feet), then turn westward, extending horizontally under the mastaba. Instead of being covered by the addition, the pit shafts were extended upward through it and overlaid with timber beams.

The builders next covered the entire structure with a *pyramid* probably having four large steps (fig. 62-4). Only the lower courses of the first step are visible, so it is not known if this construction was completed. That a four-step pyramid was for a while intended as the final form is implied because it was being cased with Tura limestone, and the foundations for a temple adjoining its north base were being prepared. In the last building stage, the four-stepper was extended to the north and west as it was greatly enlarged to the six-step pyramid we see today (fig. 62-5). The block size for the six-step pyramid was increased somewhat, averaging about 0.22 cubic meters and 500 kg (1100 pounds). The blocks were laid in alternate courses of stretchers and headers, but were not finely joined.

Fig. 63 Masonry of the first addition to the core mastaba of Djoser's pyramid.

Fig. 64 Masonry details on the west (center) base of the six-step pyramid.

Fig. 65 Level casing courses of mastaba extension (to left), and start of inclined courses of the four-step pyramid on east side of Djoser's pyramid.

Why is it proposed that the four-step pyramid was such if only the lower courses of the first step can be seen? Because the visible stones of this addition were not laid in level courses. The explanation follows.

When it was decided that the building should not continue outward, but go upward instead, a new construction technique was devised, and became a feature of all step pyramids built in Dynasty 3. Concentric walls leaning upon a central core were carried upward at an angle of 74 degrees with the horizon, a slope of seven to two. Some authors have called these "accretion walls," believing that they were sequentially added as the pyramid grew in height. But we will see in the unfinished pyramid of Sekhemkhet that all of the walls were created at the start, and rose together, the top surface at any time being more or less level. The two outermost layers stopped when they formed the first step. The next two inner layers were stopped at a height where they formed the second step, and so on. Thus each major step is comprised of two layers of walls.

Fig. 66 Exposed wall layer on north face of Djoser's Pyramid.

Notice that the masonry courses in the walls are perpendicular to the wall faces, and therefore slope inward at a 16 degree bedding angle.

Fig. 67 Step pyramid construction.

The stability gained by constructing inclined layers of masonry was aptly propounded by German-born British physicist Kurt Mendelssohn (1906-1980) in *The Riddle of the Pyramids* (1974). Not only are inward-leaning walls stabilized by their want to slide toward the interior of the pyramid, which they cannot do, but their independence allows stress relieving adjustments in response to natural forces working to settle the building.

The stepped form itself must have had significance beyond structural stability because later pyramids have stepped cores while having *level* courses and no concentric-layered interior walls.

Following Djoser we know of two large pyramids that were started but not completed, the pyramids of Sekhemkhet at Saqqara, and Khaba (tentative attribution) at Zawiet el-Aryan. Of these, the more interesting is the pyramid of Sekhemkhet. It is important because it was aborted in a condition in which we can see possible construction evidence.

SEKHEMKHET'S PYRAMID

Only a few hundred meters southwest of Djoser's pyramid is the buried pyramid of Sekhemkhet. Egyptian archaeologist Muhammad Zakaria Goneim (1905-1959) began excavating this site in 1951, and discovered the bordering *temenos* wall, the "white wall," on New Year's day 1952. Noting its similarity to that enclosing Djoser's complex, he correctly surmised that the central area would contain the remnants of a royal pyramid. He directed his workmen to dig a trial pit in that area, and "by good fortune" found the southern edge of the pyramid on February 2, 1952. Six inward-sloping layers of masonry, the sure indication of a pyramid, were uncovered before the excavating season ended in May. Because he needed time to record his findings and obtain a new grant, Goneim did not resume excavating until the following year.

In November 1953, with much excitement, Goneim's workmen uncovered the southwestern corner of the pyramid. The next year he discovered pottery jars and seal impressions in the burial chamber under the pyramid that were inscribed with the Horus name *Sekhemkhet,* "positively confirming the attribution of the pyramid to that king." Goneim placed Sekhemkhet, provisionally, as the third king of Dynasty 3, and Djoser's immediate successor. According to the Turin papyrus, Sekhemkhet reigned for six years.

Fig. 68 The entrance ramp of Sekhemkhet's pyramid.

Goneim found the superstructure of Sekhemkhet's pyramid to mimic that of Djoser. In *The Buried Pyramid* (1956), He said:

"The whole structure is 400 feet square . . . it has a maximum height of about 23 feet, but I believe that it might have been built originally to double this height and that it has been reduced by quarrying in later times. No traces of an outer casing were found."

"This is a square layer-structure consisting of probably fourteen skins of masonry which diminished in height from the centre outwards and leaned on a central nucleus at an angle varying between 71 and 75 degrees, with the beds of the courses at right angles to the facing lines. Assuming that each pair of these skins of masonry was designed to form one step, as is the case in Djoser's pyramid, we may infer that the new pyramid was intended to have seven steps in place of Djoser's six."

Sekhemkhet's average block size is about 0.23 cubic meters, or 545 kg (1200 pounds).

Goneim discovered construction evidence that could be pertinent to the building of later pyramids:

> But on three sides of the structure I found traces of what are almost certainly construction embankments – the "foot-hold embankments" described above, which were intended to give the workmen access to the higher portions of the monument.
>
> The top of these embankments is higher than the present level of the pyramid, suggesting that it was originally built up to a higher level than the present one, and that the upper courses have been removed during quarrying operations.
>
> Also, on the west side of the structure, nearest the quarry, we found part of the original construction ramp up which the stones were hauled.

I don't think Goneim's suggestion that because the embankments were higher than the pyramid its original height must have been reduced by quarrying is compelling. If the pyramid later served as a quarry why would the defilers have bothered to bury it? Perhaps the western embankment was higher than the outer walls, and crossed the interior of the pyramid, because that was the easiest way to get blocks to the far side of the structure.

Fig. 69 Construction ramp and embankment on west side of Sekhemkhet's pyramid. Drawn from Goneim (1956, plate 25).

Goneim died in January 1959. I don't know if any more work has been done at this site. Further excavation may tell us more about the large western ramp and surrounding embankments. The title of Goneim's book is, nonetheless, fitting. Sekhemkhet's pyramid is indeed buried. But we do not know if the material surrounding the pyramid is construction-related or camouflage. I'll have more to say about Sekhemkhet's pyramid in chapter ten.

5

MEIDUM

THE MEIDUM PYRAMID

The next large pyramid was erected fifty kilometers (31 miles) to the south, at Meidum (MY-*dum*).

Fig. 70 The Meidum Pyramid, from northwest.

From a distance the building looks like a squat tower on a hill curiously out-of-place on a flat desert. The picture is deceiving. The hill conceals the base of the first *regular* pyramid, a structure with four flat triangular faces coming to a point at the top. A mixture of wind-blown sand and material fallen from the upper portion of the pyramid surrounds much of its base. In 1978, only on the middle portions of the east and north sides could one see faces of the regular pyramid form.

How the pyramid fell to ruin is controversial. I'll get to that in a moment.

In this monument we have three distinct projects. In order of construction the projects were labelled E1, E2, and E3 by Ludwig Borchardt (1863-1938), the German architect who first described the evolution of the building. Petrie and Wainwright had previously discovered its basic structure.

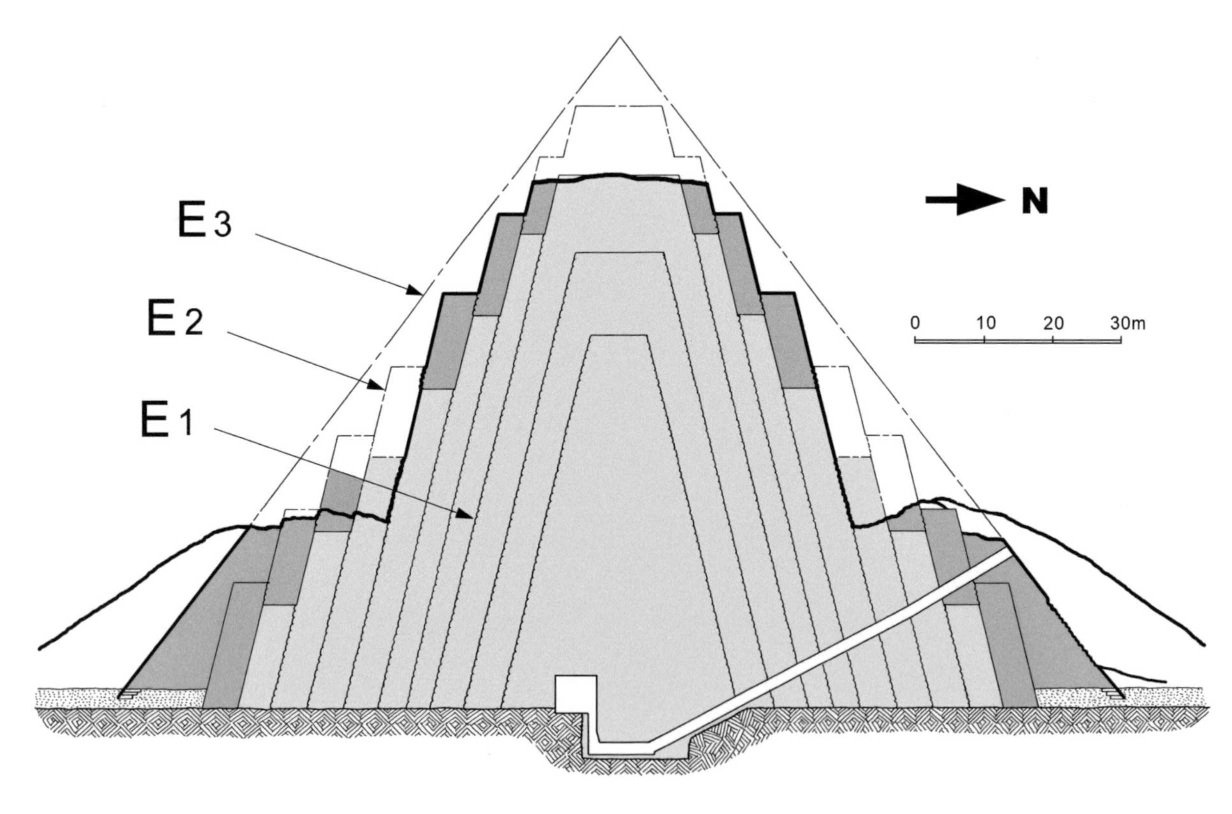

Fig. 71 Section through the Meidum pyramid.
Redrawn from Maragioglio & Rinaldi (1964, vol. III, Tav. 3)

Project E1 resulted in a step pyramid of seven tiers. Its blocks are half again larger than those of Sekhemkhet. Because the outer masonry is smoothly dressed, and provision for a door was made at its entrance, E1 may have been considered, for a while, the final form. We do not know, however, if its seventh step was completed.

Fig. 72 The Meidum pyramid, SW corner of E1 construction.

The pyramid was enlarged in project E2 with a coating of masonry which completely covered E1, making it an eight-step pyramid. In applying the E2 layer the builders followed the same practice as in E1 - inclined buttress walls of inward-sloping courses. The construction is similar to that of Djoser, but at Meidum each step consists of one layer, not two. Average block weight is about 770 kg (1700 lbs).

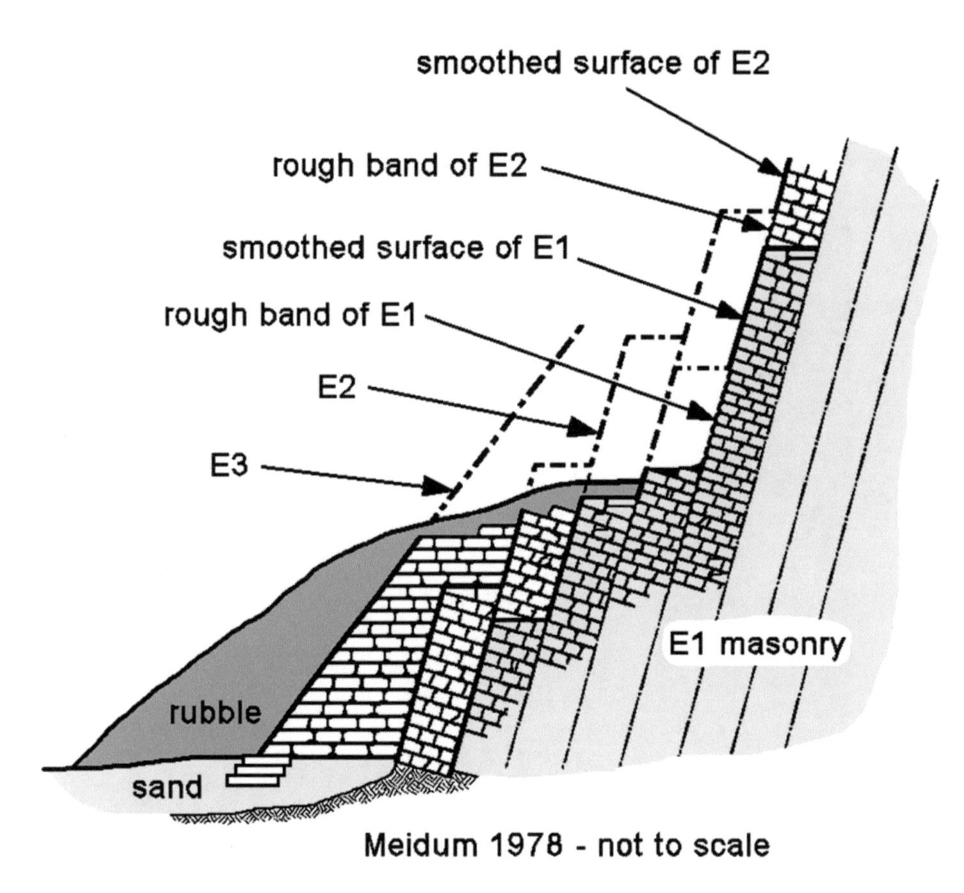

Fig. 73 Masonry of the Meidum Pyramid.

The casing blocks of E1 and E2 are much better fitted than those of previous pyramids. But blocks behind the casing are not well squared, tightly packed, or even in distinct courses. This rough-work can be seen in the large hole in the north face of the pyramid (figure 74) and in even larger holes on the west and south faces. The result of this apparent sloppiness will be presented shortly.

Fig. 74 Roughly-shaped packing blocks in north face of Meidum Pyramid.

The face of E2, like E1, is dressed smooth. Where the entrance passage enters the E2 addition there are door-hinge cut-outs in its walls, evidence of a possible closing. Like E1, this suggests that E2 was thought to be the end of the job.

Not so!

One more coating was added, this time filling in the steps of E2 and converting the building to a regular pyramid, project E3.

The masonry of E3 is similar to that of E1 and E2 in having well-fitted casing stones and ill-fitted packing stones, but it is quite different in two other respects. The main difference is that for E3 the masons placed both casing and backing stones in *level courses*, not inward-leaning. The second difference is that the foundation stones for the first course of masonry lie on a bed of *sand* rather than bedrock. The divergence of building methods displayed in E3 supports the contention of Maragioglio & Rinaldi that E3 was constructed under different supervision than the first two projects.

The Meidum Pyramid is now attributed to Sneferu, the first king of Dynasty 4. This is based on a New Kingdom graffito found by Petrie and Rowe in the chapel adjoining the eastern base of the pyramid, and the names of some of Sneferu's officials found in nearby tombs. Note, however, that the chapel contains two stelae that were never inscribed with the name of the pyramid's owner. Thus the pyramid may have been abandoned before it was completed (or dedicated).

Some have suggested that only project E3 was completed by Sneferu. They think he commandeered for his own use the building E1/E2 erected by his predecessor, Huni. Huni's reign was long enough (24 years – Turin papyrus) that he could have produced a large monument. But Huni's name has not been found in any nearby tomb or temple, so the only evidence we have points to Sneferu as builder of all three stages.

The Meidum quarries

According to the Klemms, except for a portion of the outer casing of E3, all stones of the pyramid, including the casing of E1 and E2, are the same material, a good grade of local limestone. They say that there is a "possible, but not entirely convincing quarry site about some 600 m south of the pyramid and . . . an extended quarry area about 1 km south of the pyramid." They offer that although these local stones "dominate the unfinished E3 phase masonry," the supply of good local stone may have been exhausted, requiring the masons to resort to stone that "greatly resembles the Maasara limestone."

The broad shallow grooves

On the vertical faces of the fifth and sixth steps of the E2 construction, on the eastern side of the pyramid, broad shallow grooves are sunk into the casing.

Petrie said the upper groove is about 211 inches wide, and the lower (fig. 75) about 195 inches wide. His estimate that both are about two to three inches deep appears right to me. Petrie thought these grooves "were analogous to the grooves on the successive coats of brick mastabas, indicating where the false door and *ka* chamber [ediface containing a statue representing the king's external appearance] lay behind them in the first body of the mastaba; hence these grooves might indicate that there was a *ka* chamber in the first body of Sneferu's mastaba." To me these grooves seem too expansive to have served as false doors.

Fig. 75 Broad shallow band in face of 5th step of the E2 project.

Dieter Arnold believes the grooves may have stabilized the sides of a masonry ramp that rose from the "approach road" (described next) that Gerald Wainwright excavated in 1909-1910. Yet he admits that even a masonry ramp of this size would not have had vertical sides. I'll present my own conjecture on these grooves in chapter ten.

The approach roads

The eastern road lies south of, and nearly parallel to, the ceremonial causeway that comes up from the valley and joins the chapel at the east base of the pyramid. Some writers believe it was part of a ramp used to bring stones to the upper portion of the pyramid.

But Maragioglio & Rinaldi pointed out (1964, 50), that because this road sloped at only 6.3 degrees, "The supposed ramp did not therefore reach the pyramid near the top, but only at the level of the second step, and accordingly could not fulfil the purpose of a working ramp, that is to say the transport of the blocks up the summit of the monument. We think, therefore, that the approach was nothing but a road to reach the plateau, and that it was abandoned before the beginning of the actual building work."

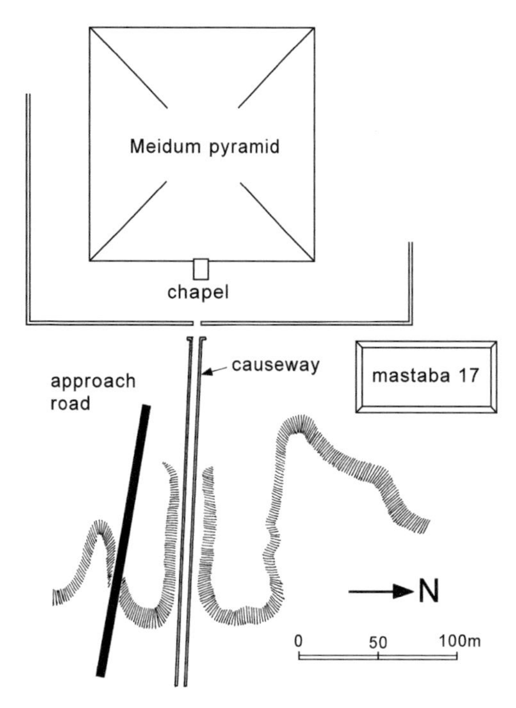

Fig. 76 The Meidum Pyramid, causeway, chapel and eastern approach road. After Petrie, et al (1910).

Arnold (1991, 82) described a similar "ramp" on the south side of the same pyramid:

"Three hundred meters from the foot of the pyramid another ramp was found. It was only 3.25 meters wide, made of parallel brick walls, with interior filled with builders' debris, and exactly with its meeting point with the pyramid casing on the sixth step, a vertical groove is visible again, about 3.50 meters distant from the corner."

I agree with Maragioglio & Rinaldi that the eastern road (and southern also) was not used to raise blocks on the pyramid. I think it's probable that both roads were used, during one building phase or another, to bring materials to the pyramid.

On the idea that large ramps were used to build the Meidum Pyramid, Maragioglio & Rinaldi said (1964, 38):

> The Meidum Pyramid, with its different stages of construction, appears to us to be a most convincing evidence against the currently accepted hypothesis that the construction of the pyramid was carried out by means of long working ramps, either of the kind imagined by Borchardt or of that conjectured by Lauer. In fact, the succession of the stages of construction would have involved, in the case of three stages, the erection and subsequent destruction of no less than three working ramps . . . which, it must be noted, should have been as high as the pyramid itself (or only a little less) and therefore really enormous. Since the stages subsequent to the first should also have begun at the bottom, the preceding ramps were thus no longer useful but constituted an obstacle for the work and should have been removed, at least in part, prior to the enlargement and casing of the pyramid. All this would have required so much labour that we do not think this idea to be acceptable.

In addition to the argument Maragioglio & Rinaldi made, if a large ramp (or ramps) served to raise the Meidum pyramid, where is the ramp material now? The desert around the pyramid is flat. In cleaning up after completing the pyramid, assuming it was completed, did the builders spread the ramp material so thin on the ground that no trace of it can be noticed today? This is another question I'll attempt to answer in Chapter ten.

Mendelssohn's collapse theory

In *Riddle of the Pyramids,* Mendelssohn proposed that the poor foundation of E3 allowed subsidence of its casing. Further, the roughly-made packing stones produced lateral (sideways acting) forces which resulted in an unstable structure. A catastrophic collapse of the pyramid occurred, perhaps triggered by a heavy rainstorm. The collapse occurred during the last stage of construction, resulting in abandonment of the site.

Mendelssohn argued that the Meidum disaster forced a design change in the next pyramid, already half built at Dahshur, the change that caused the Bent Pyramid (next chapter) to be "bent."

Lastly, Mendelssohn supposed that pyramid construction, in fulfilling the purpose of civil service make-work, at times entailed labor on more than one pyramid at a time.

Mendelssohn's ideas are attractive. His explanation for the unusual shape of the Bent Pyramid is at least plausible. Further, he is probably right that the builders often started the next pyramid before the current one was completed. But his Meidum scenario contains a few dubious conclusions. One of these is his proposal that a sighting pole was erected on top of the stepped core of every pyramid to guide accurate placement of casing for the regular pyramid form. In chapter nine I will describe what I believe is a more probable alternative to his sighting pole.

Another disputable conjecture concerns the completeness (or not) of the building stages. Mendelssohn argued:

a. The top portion of the enlarged step pyramid E2 was never completed.
b. Therefore E3 was started before E2 was completed.
c. E3 could not have surpassed the level of E2, so E3 likewise was not completed.

This argument begins with a shaky premise. Mendelssohn accepted the opinion of M.A. Robert, a French surveyor who scaled the pyramid in 1899. To aid his survey of the Fayum region, Robert was setting up a marker flag on the highest point he could find. He reported (1902) that the highest step was easily accessible from the east side and "seems to have never been completed."

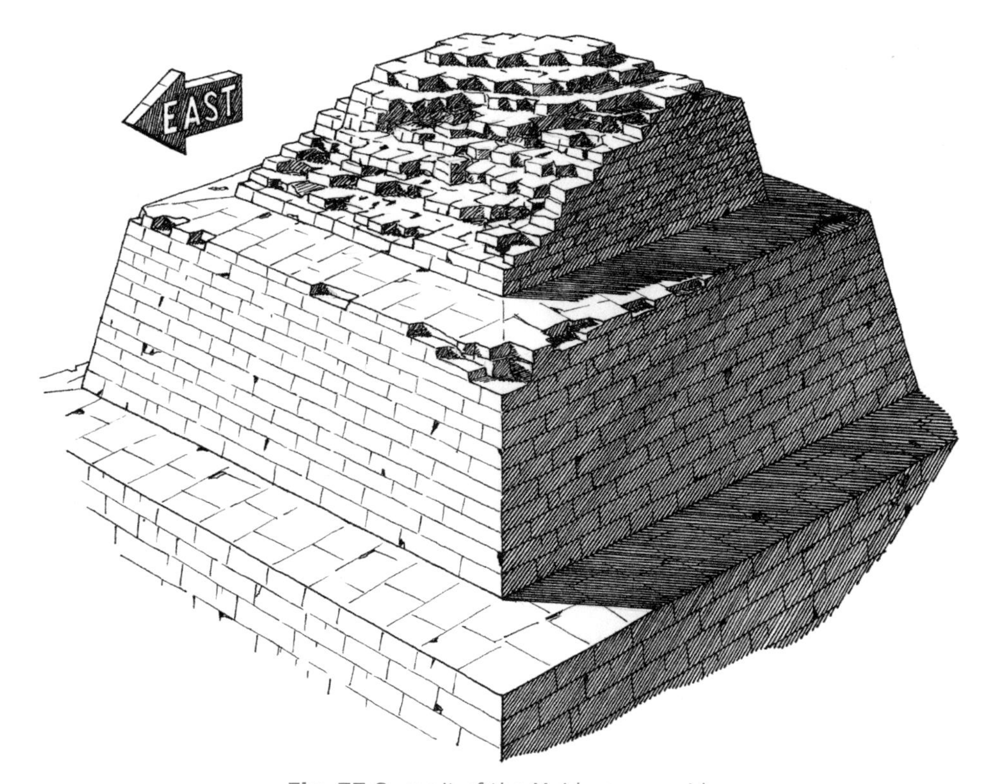

Fig. 77 Summit of the Meidum pyramid.

The unfinished appearance of the summit may have another explanation. The top surface is not only irregular but slopes down to the east. The typical aspect of unfinished pyramids is, however, nearly level-topped. Why? Because a continuously-leveled perimeter is necessary in order to maintain accuracy of the building. To explain the down-east slope of the Meidum summit, we should consider that blocks may have been removed and tumbled down the east face for use in later projects.

Completed or not, the pyramid became ruined, and so I will venture a modification of Mendelssohn's conjecture:

As Sneferu began the E3 addition that would transform the building into the first regular-right-square pyramid he decided to build a more impressive monument at Dahshur, 42 kilometers to the north. He left a small crew at Meidum to work the E3 experiment.

The new building at Dahshur (ultimately to become the Bent Pyramid) would be the first designed at the start as a geometrically regular pyramid. Its base length would be 300 cubits. It would have steeper sides, with a slope of 7/4 (60 degrees), and rise to 260 cubits, much higher than the 175 cubits of the Meidum E3 building.

Work progressed on the new pyramid, but as its height reached perhaps forty meters, signs of stress appeared as vertical cracks in its outer casing. Sneferu's master builder was alarmed. He decided to add a buttressing mantel to the pyramid, a coating that increased its base length by sixty cubits. Its new face would be lowered to a slope of 7/5 (angle 54° 27'). The evidence:

In the northern and western corridors of the Bent Pyramid, at distances from the outer face of the pyramid of 12.6 meters (41.3 feet) and 10.6 meters (34.8 feet) respectively, there are continuous joints in the masonry which form planes roughly perpendicular to the passage axes. I indicate these points as A and B in figure 78. By "continuous joints" I mean joints which define a continuous plane in the ceiling, floor, and walls. This condition is unusual because Egyptian masons typically broke joints between layers of masonry to create stronger bonding. Just to the interior of the continuous joints are cuttings in the walls suggesting (temporary) wooden doors were once installed at these places.

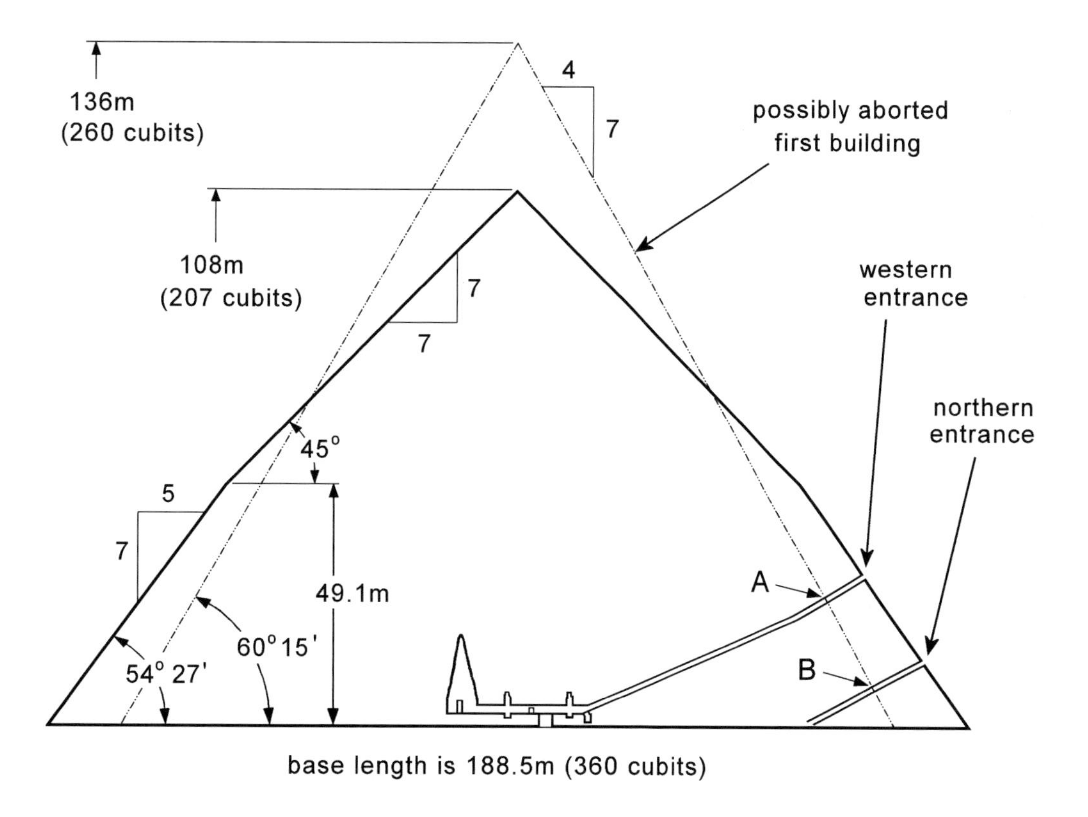

Fig. 78 Maragioglio & Rinaldi's proposed steep-faced pyramid.

The similarity of these probable temporary closings induced Maragioglio & Rinaldi to draw both corridors in the same view. They noted that a line connecting the continuous joints formed an angle of about sixty degrees with the horizon. From this they hypothesized that a smaller but steeper pyramid had been under construction. When signs of crushing appeared, the architect relinquished the steep pyramid in favor of a larger building with lesser slope.

As Meidum E3 was nearing completion an earthquake occurred. The building was gravely damaged, but did not collapse. Its mantle bulged in several places. Casing stones were lost, creating large rifts near mid-level of the structure. The horrified builders knew that such injury was irreparable. Having learned much from the enterprise, they picked up their tools and moved north to join their fellows at Dahshur.

The same tremor that shook the Meidum Pyramid made its mark at Dahshur. Here the newly-buttressed pyramid had reached a height of forty-nine meters (161 feet), the present "bend" level. One can see, in the north descending corridor, at the continuous joint I mentioned above, signs of slippage of the outer masonry with respect to the inner.

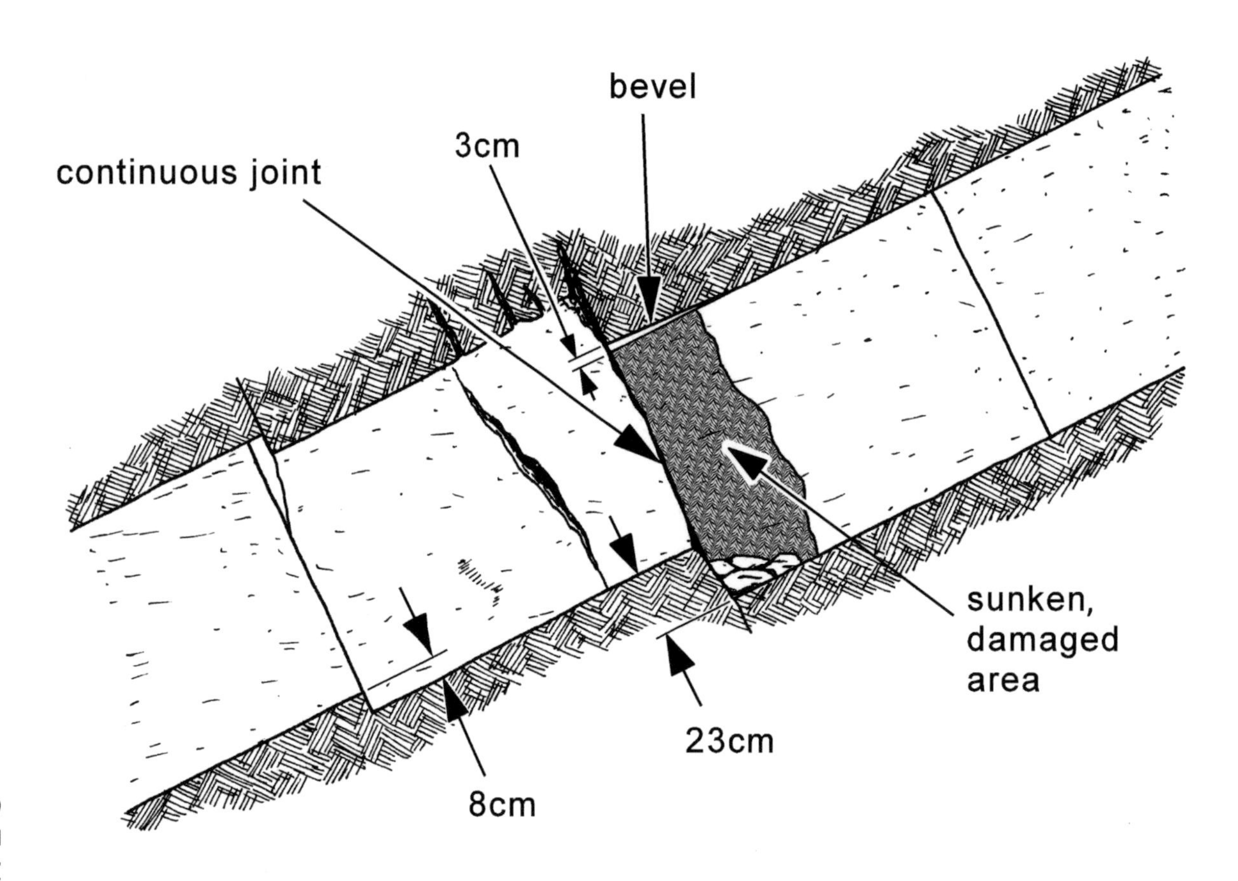

Fig. 79 Discontinuities in the north descending corridor of the Bent Pyramid. Adapted from Maragioglio & Rinaldi (1964, Tav. 11).

The slippage is evidenced by a three-centimeter vertical dislocation of the northern portion of the shaft, a displacement that was repaired by beveling the ceiling block that had shifted downward. This apparently cosmetic repair was certainly made while the pyramid was still open, i.e., by the builders. This damage corresponded with the Meidum injury, resulting either from the same tremor or another, more local, event.

Fig. 80 The beveled repair of the ceiling in the northern entrance of the Bent Pyramid.

A second settlement produced a further twenty-centimeter displacement at the same joint, and a shift at the next joint down the passageway of eight centimeters. These injuries were not repaired. The second event, more violent, and probably occurring years after the first, may have caused a substantial volume of material to be lost from the upper portion of the Meidum pyramid.

When did the Meidum Pyramid suffer its greatest damage? During Petrie's 1891 clearance of the east face he uncovered the small temple that contains the uninscribed stelae I mentioned before. In the temple he noted graffiti which date to Dynasties 19 and 20, indicating it was not buried in these times. He also found intrusive burials of Dynasty 22 not far below the surface of the rubble that surrounds the pyramid. From these facts it follows that the pyramid became most damaged sometime between Dynasties 20 and 22 (1200-800 BCE).

Later the ruined building was used as a quarry. A substantial amount of material has been removed from the base of the present "tower." If this were not so we would see a talus of rubble at the base of the tower instead of the hollowed-out aspect now evident.

The pi angle

Part of the outer casing of E3 is accessible on its north and east sides. The smoothed casing makes an angle of about 52 degrees with the horizon. This slope occurs in later pyramids, notably Khufu's Great Pyramid at Giza.

Some folks think that arcane information was incorporated into the pyramids, especially the Great Pyramid. They believe that the builders intended the Great Pyramid to model certain measures of our planet. Specifically, they believe that the builders made its perimeter equal to one half arc-minute of longitude at Earth's equator, and its height equal to its perimeter divided by 2π.

Let's see how close the builders came. A *geographical mile* is defined as one arc-minute along Earth's equator, a length of 1855.4 meters, or 6,087 feet. According to Cole's survey (1925), the perimeter of the Great Pyramid is 921.456 meters, or 3023.15 feet. Twice 3,023 feet is 6,046 feet, which is close to the 6,087 feet of the geographical mile. If we assume that the Great Pyramid's original height was designed to be its perimeter divided by 2π, its height would have been $3{,}023.15/2\pi$, or 481.15 feet. And since 481 feet (also 280 cubits) is the height that many Egyptologists believe was the original height of the Great Pyramid, the numerologists' proposal could be right.

But countering the arcane model are the following points.

First, there is no evidence that the ancient Egyptians conceived the idea of polar coordinates (angles). And even if they did, our system of 360 degrees in a circle, with 60 minutes to a degree, is arbitrary. Someone else's circle could just as easily have 1000 degrees. In the Rhind mathematical papyrus, which dates to Dynasty 25, the writer uses X-Y values (Cartesian coordinates) in describing pyramid slopes. Cartesian coordinates were also used throughout ancient Egypt for laying out arches and measuring portions of land. Thus I think it is unlikely that the architect(s) of the Great Pyramid used arc-minutes along the equator in his calculations.

Second, there is no evidence that the ancient Egyptians were acquainted with the transcendental number 3.1415926 . . ., or "pi". In the same papyrus the writer calculates the area of a circle by squaring 8/9ths of its diameter (in error, incidentally, by only a half percent). If the pyramid builders were acquainted with π, why didn't they use it to calculate areas or circumferences of circles?

Pyramid slopes are called *sekeds* in the Rhind mathematical papyrus. Only one number, the run (or x value), is given for any slope. The rise (or y value) is understood to be one cubit, or seven hands. Thus a seked of 5 ½ means a slope of seven to five-and-a-half, or 14/11, which is almost certainly the slope of many pyramids, including Meidum, Khufu's, Menkaure's, and at least two of the small pyramids on the east side of the Great Pyramid.

Now compare the heights of the Great Pyramid using the two methods just described. Let's accept Cole's 755.79 feet as the base length of the pyramid. If we use the method favored by numerologists the height of the pyramid would be, as noted earlier, 481.15 feet. The slope of the apothem would be 51° 51' 14". If we use the slope of 14/11 we multiply half the base length (377.89) by 14/11, yielding a height of 480.96 feet. The slope of the apothem would be 51° 50' 34". The difference between these two methods is miniscule, only 5.8 cm (2.3 inches) in height!

Given these virtually identical values, I think it's probable that the builders used seked proportions for all the pyramids. The process by which the first regular pyramid was born supports this contention. As J.P. Lauer noted (1974, 307), the Meidum builders covered a step pyramid with sloped surfaces that came closest to matching the planes that connected the outer edges of the steps of E2, yielding a gradient of 14/11.

Fig. 81 Greg measures the casing angle on the north face of the Meidum pyramid.

I doubt this issue will ever be settled. But one should be suspicious of pyramid slopes reported by modern scholars in units of degrees, minutes, and seconds. These values imply a level of precision impossible to verify. Even when presented as average readings (an unspoken admission of imprecision) the spread of one's measurements, along with an estimation of the accuracy of the measuring instrument, should be provided.

When I measured pyramid slopes, including Khufu's, with my clinometer (+/- 10' estimated accuracy) I obtained values on a series of adjacent blocks that differed by up to two degrees. On any one block I found at least one degree variation, so casing blocks cannot give us the designed slope of the pyramid. I doubt that surveying techniques offer much improvement. In 4,500 years it is likely that some amount of settlement has occurred. If we could ignore settlement, the best datum for measuring the height of a pyramid would be sighting its apex, but no pyramid retains a capstone.

Capstones might give us the most reliable indication of the intended slope of a pyramid. Unfortunately, only one has been recovered that is both finished and intact. This beautiful stone, of polished black basalt, is from the pyramid of Amenemhat III at Dahshur. Its slope is 54 degrees 27 minutes, same as the lower portion of the Bent Pyramid.

Fig. 82 The pyramidion of Amenemhat III's pyramid, Cairo Museum.

6

DAHSHUR

THE BENT PYRAMID

Its ancient name was *Southern Pyramid Kha of Sneferu*, but its unique shape inspired its popular name, the Bent Pyramid. The double-slope feature was once thought to have been induced by the builders' desire to either finish the pyramid in haste or represent some kind of duality. Today most Egyptologists believe that an alteration of the original design was made for engineering reasons. Lauer, Badawy, Maragioglio & Rinaldi, and Lehner think the designed height of the building was reduced when signs of stress (cracks, flaking) in the casing and internal apartments appeared during construction. Also, one can infer from the pattern and extent of missing casing that the building has been subjected to violent shaking.

Fig. 83 The Bent Pyramid, from NNW.

I explained why I think signs of stress compelled the Dahshur architect to add the buttressing mantle of lower slope to the pyramid. But to force a second slope change, that of the top portion, he would have needed a stimulus as dramatic as the first. The cracks in the casing, and settlement in the north descending passage, along with (possibly) concurrent damage to the Meidum pyramid, would seem dramatic enough. Of course we cannot know if the apparent shaking damage evidenced at the Bent and Meidum pyramids resulted from simultaneous events.

The Bent Pyramid rests on a surface of rock and compacted clay. Its casing stones are large and well fitted. Most are laid as headers. Many backing stones, like the casing, are Tura limestone. Often two casing stones are stacked to gain the required course height. The exposed core blocks vary considerably in size, and are not tightly packed. Unlike the masonry at Giza, large gaps appear between the core blocks.

Fig. 84 Gaps in interior masonry of the Bent Pyramid, west face. Note the variability of block size and the human figure at lower right.

Fig. 85 Casing and backing stones at NW angle of the Bent Pyramid (looking east).

Bedding seats of casing blocks in the lower section (up to the bend level) slope inward at five to eight degrees (Maragioglio & Rinaldi vol 3, 1964, 56). Casing courses of the upper section slope inward at about three-and-a-half degrees (Vyse, 1842, 66).

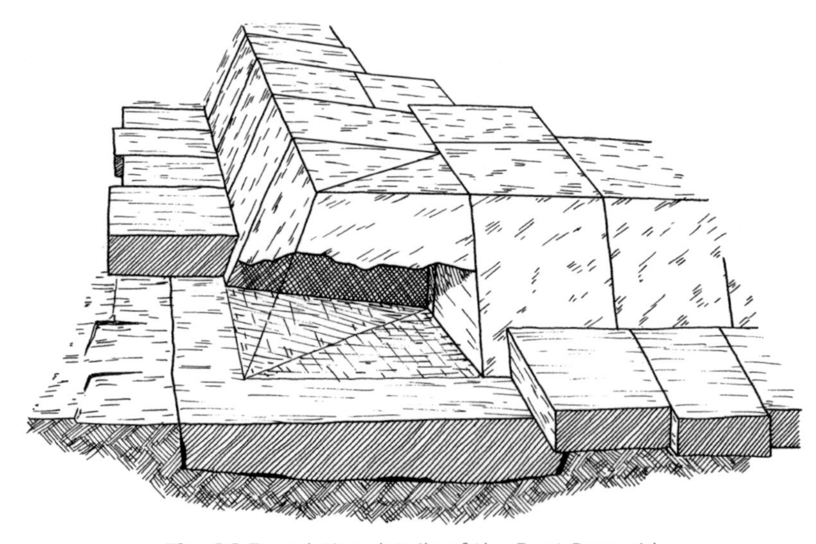

Fig. 86 Foundation details of the Bent Pyramid.

Was the Dynasty 3 technique of using inward sloping courses selected as a result of Meidum E3 blunders? I don't think so. If the architect knew of a weakness in the Meidum building, I doubt he would have commenced the buttressed configuration of the Bent Pyramid at a slope steeper yet.

The seked of the lower portion of the Bent Pyramid is 5, a slope of 7/5. The slope of the upper section is not known exactly. The last survey was made in 1952 by Hassan Mustapha while working for Egyptian archaeologist Ahmed Fakhry (1905-1973). Mustapha could not measure the base of the pyramid because the corners are missing. Instead he used a theodolite and trigonometry to determine points on the pyramid edges. Since the apex of the pyramid is also missing he was forced to guess its position. He reported that the upper portion had an inclination of 43 degrees 21 minutes (Fakhry, 1959, 67).

I think it's likely the upper section of the Bent Pyramid was built with a seked of 7 (45 degrees), and settlement over 4,600 years has reduced its height. The subsidiary pyramid on the south side of the Bent Pyramid has this slope, as does the next large pyramid Sneferu built, popularly called the Red Pyramid.

Fig. 87 The subsidiary pyramid (100 cubit base) on the south side of the Bent Pyramid, looking west-southwest. Seked = 7 (45 degrees). No nucleus structure is apparent.

The upper section of the Bent Pyramid is cased with smaller blocks, on average, than the lower. Some writers have claimed that the quality of materials and workmanship is inferior to that employed in the lower portion. But Maragioglio & Rinaldi said that "the difference in quality and work is not so remarkable", pointing to the "general characteristic [of the pyramids] that stones progressively diminish in size from the base upwards."

Fig. 88 The author descending from north entrance of the Bent Pyramid. Photo by Sheri Moores.

Of all pyramids, this one competes only with Khufu's in having the most intricate internal features. It is the only pyramid with two entrances, each leading to a separate chamber (the two entrances of Khafre's pyramid lead to the same place). The usual northern entrance descends to a tall, narrow antechamber (fig. 89-A) that adjoins a room (fig. 89-B) having a lofty corbelled ceiling.

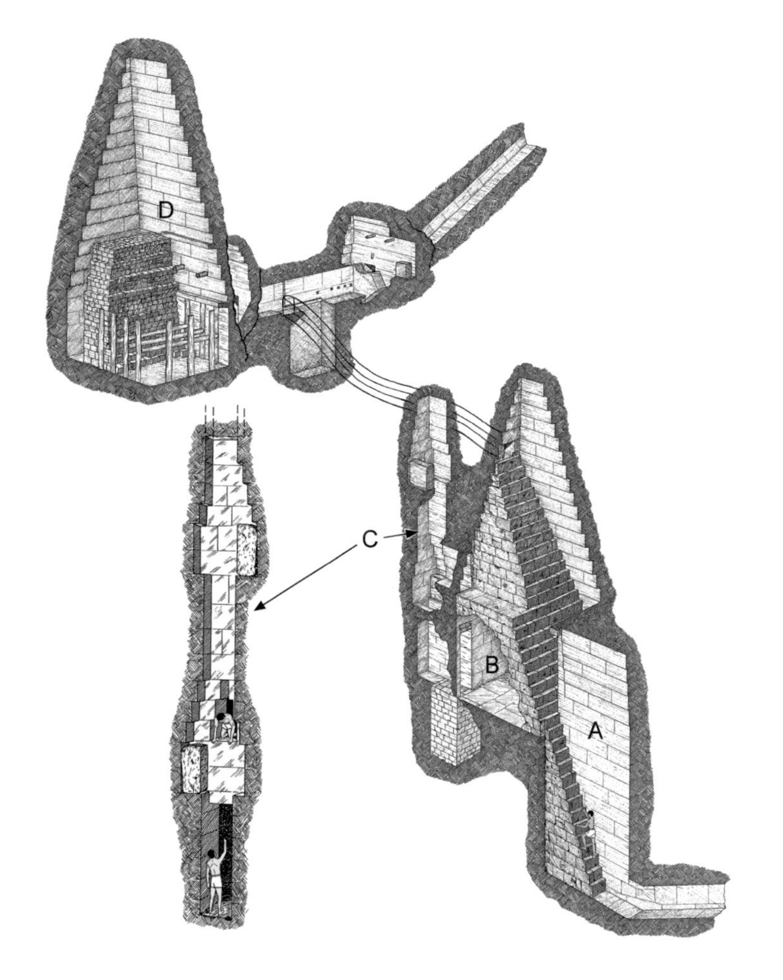

Fig. 89 Interior of the Bent Pyramid. Computer-aided perspective drawing by author using orthogonal measurements from Maragioglio & Rinaldi (1964, vol. III, Tav. 11 & 12). The stairways shown in rooms A and B were temporary and probably removed by the builders.

The second entrance descends from the western face to a room (fig. 89-D) similar to room **B**. There are reasons why the western system may have been planned for the actual burial of the king. First, the atypical location of the western entrance would not have been foreseen by tomb robbers, especially considering that the northern entrance was made obvious by a large architrave above its position. Second, the western corridor is protected by two portcullis gates. That the northern passage was not similarly defended suggests it may have been planned as a false lead.

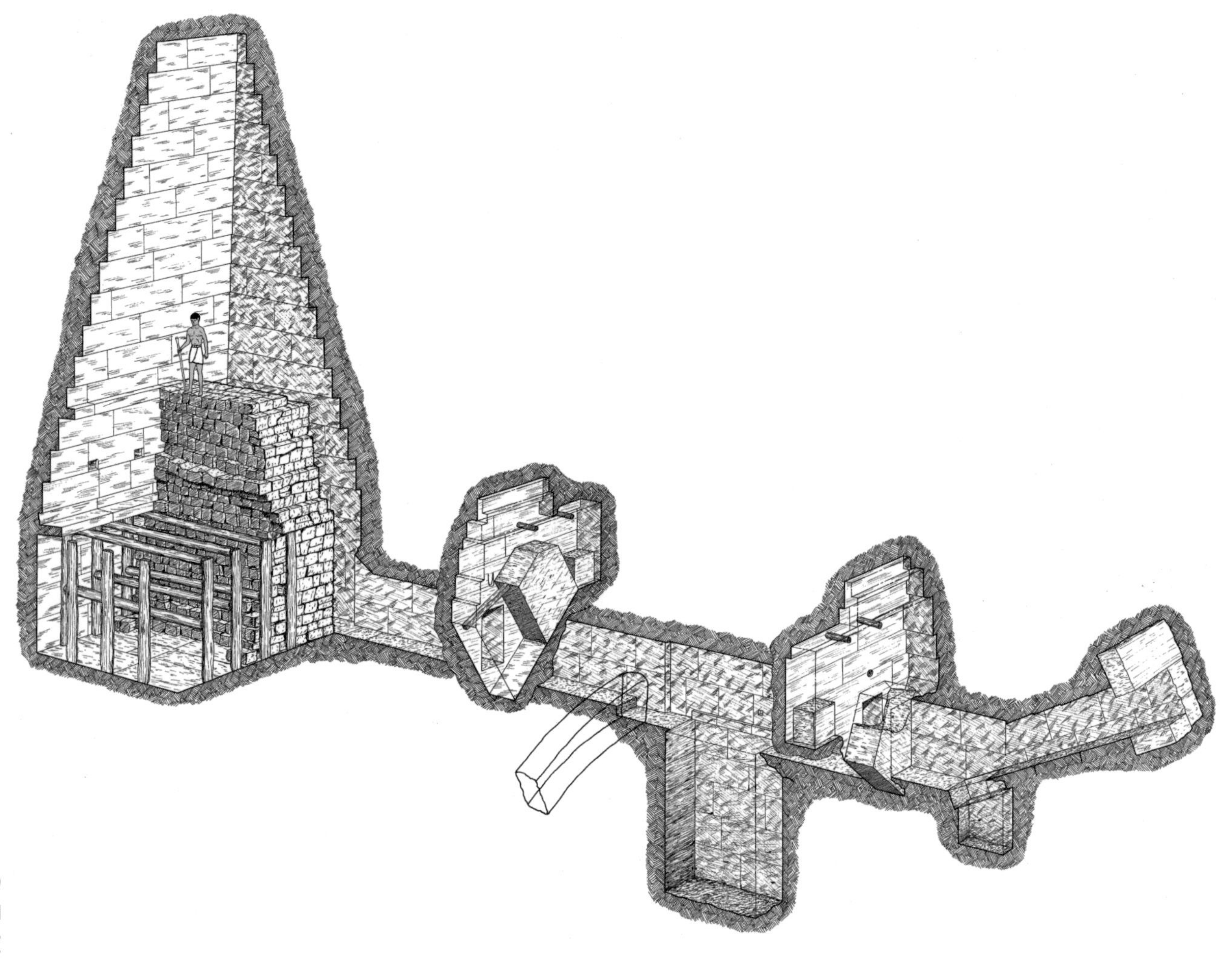

Fig. 90 The upper crypt and corridors in the Bent Pyramid.

Evidence suggests that the inner compartments were in the following condition when the builders ceased work:

A stairway of small limestone blocks, mortared together, rose from the base of antechamber **A** to a point high on the south side of the ceiling of room **B**. Traces of mortar on the walls of the antechamber and room **B** delineate the stairway. At the top of the stairway a rough-cut tunnel wound its way southward to the horizontal portion of the western passageway, intersecting it between the two portcullis gates.

The western portcullis had been lowered and sealed with plaster on both sides.

The entire sloped portion of the western corridor was filled with limestone plug-blocks individually sealed with plaster on their outer (western) faces. The last plug was a casing stone unobtrusive on the pyramid exterior.

Though the walls and ceiling of room **B** were nicely dressed, the same was not done in room **D**. Except for a steep, narrow staircase at its doorway, room **D** was filled with small limestone blocks to at least the level where its corbelled ceiling begins. Fakhry removed many of these blocks in 1946. As he did, he uncovered a framework of cedar logs. The eastern portcullis was not closed. It remains in standby position, propped by a wooden beam, suggesting that this gate never protected a burial.

How then do we explain the state of the room most likely planned for the king's eternal abode? The filling blocks must be original. Fakhry found Sneferu's name on one. Perhaps the king wanted a large burial vault, but proofed against collapse. The cedar framework gave delineation to the room that would otherwise be lost. The king's *ka* did not need open-air chambers. It was important that rooms were present, stone-filled or not. When it came time to inter the king, his coffin could have been inserted into the filling or placed on top. Another explanation may be that the logs and small blocks served as temporary support for the walls of the chamber while its ceiling was under construction. The supports would have been removed if it was to actually serve as a crypt.

At some point the builders decided that the western entry was superfluous; they devised a trickier plan for interring the king. They dug the winding tunnel from north to south, making room **D** accessible from the northern system. They lowered the outer (western) portcullis and completely sealed the corridor. The new plan was almost perfect for concealing room **D**. If the builders had covered the tunnel entrance in the ceiling of room B, and removed all traces of the stairway leading to the tunnel, it is doubtful that room **D** would have been discovered. Why wasn't the concealment plan for room **D** realized? Because Sneferu decided to build a bigger and better pyramid.

The chimney

Accessible through two openings in the southern wall of room **B** is a feature called the "chimney" (fig. 89-C). From the pavement of room B the chimney rises slightly more than 15 meters (49 feet). There are two widenings in the shaft where standing limestone slabs could be tipped over, closing the passage. The lower slab remains standing; the upper is missing. We cannot be sure that the ceiling of the chimney, two slabs side-by-side, are not similar gates blocking continuation of the shaft.

The chimney may have served the same function Lehner proposed for the "Well" shaft in Khufu's pyramid, a control datum for guiding the construction of the interior apartments. The chimney would be well-suited for this purpose, lying as it does "almost exactly on the central vertical axis of the pyramid," (Maragioglio & Rinaldi).

The portcullis gates

The Bent Pyramid is the first to contain a portcullis blockage, in this case two executions of the same elegant design. Both are in the western passage. Each consists of a limestone slab resting on a surface that slants at about 35 degrees. The western portcullis was closed. The eastern is propped open with a wooden beam. The prop looks like a guard against accidental tripping of the primary mechanism, a transverse wooden beam holding the sliding gate in the standby (open) position. One end of the beam is in a shallow hole; the other in a slot such that a rope could be used to pull it up, thereby releasing the gate and blocking the passage.

From table 3 you can see that the angle of friction for a limestone block on a limestone surface is about 26 degrees. Thus, these two portcullis slabs, lying on surfaces which slope at 35 degrees, would have slid into place without assistance. At the same time, the angle was not so great as to put too much pressure on the transverse beams that held them back – a clever design!

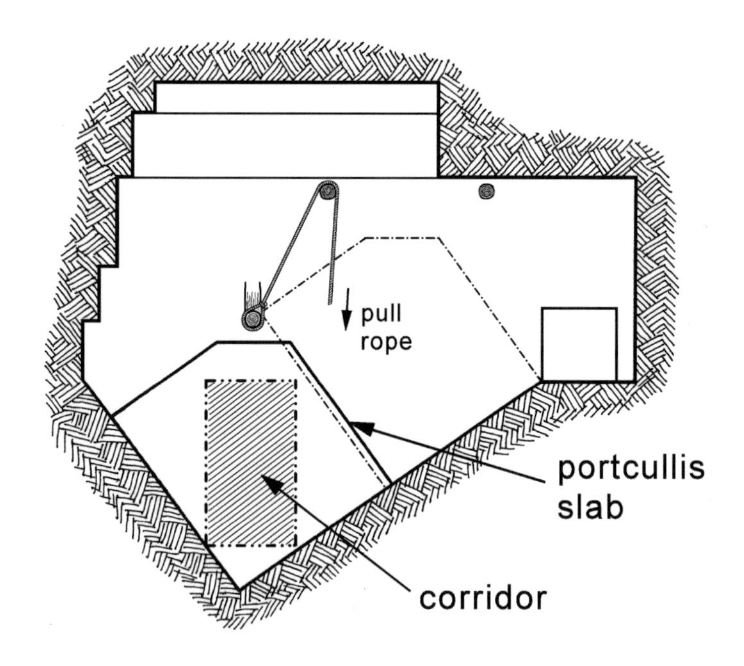

Fig. 91 Portcullis design in the Bent Pyramid. From Maragioglio & Rinaldi (1964, vol. III, Tav. 13).

I can't leave the Bent Pyramid without mentioning the mystery of the howling wind. In *The Monuments of Sneferu at Dahshur,* Fakhry reported:

"On some windy days, there can be heard inside the pyramid, and especially in the horizontal corridor between the two portculli at the end of the ramp of the western entrance, a noise which continues sometimes for almost ten seconds. This occurred several times and the only explanation is that there is still some communication with the exterior, and probably an undiscovered part of the interior of this pyramid still exists. We should always bear in mind what Perring told us about the sudden great noise and the very strong current of air which continued for two days and then stopped suddenly. This means only one thing: there is some connection with the outside. The western entrance must be ruled out because it was completely filled with blocks, one after the other, and every block was plastered all around it, and the last block in it was one of the stone casing. The exit which caused this strong current of air in 1839 and which still causes noises heard inside the pyramid before the opening of the western entrance in 1951 is still to be located."

Apparently we still have things to learn about this enigmatic building.

THE RED PYRAMID

The *Northern Stone Pyramid of Dahshur* lies two kilometers north of the Bent Pyramid. It is also called the Red Pyramid because of the reddish-tinted backing stones that were made visible when most of its casing was stripped away.

Fig. 92 The Red Pyramid.

Egyptologists agree that the Bent and Red pyramids were constructed for Sneferu. If the Meidum Pyramid was Sneferu's as well, then Sneferu directed the construction of three large pyramids.

The Klemms found that the quarries from which the core stones were obtained lie southwest of the pyramid. Two quarry roads, or drag ramps, lead from these quarries to the pyramid. I mentioned these in chapter three (figure 39).

On a piece of casing at the base of the pyramid, German archaeologist Rainer Stadelmann (1933–present) found an inscription (85, 100) that said it was placed "in the 15th counting," which would be the 29th year of Sneferu's reign. Stadelmann found an inscription on another casing stone, thirty courses higher, that is dated only four years later. At the level of thirty courses more than half of the volume of the pyramid was established. Thus it is difficult to believe that the builders could complete half the volume of the pyramid in only four years. By my rough calculations it took about eighteen years to build this pyramid completely (see section on pyramid construction time in chapter eight).

The Red Pyramid is not small. Were it hollow it could contain the Bent Pyramid. Its base length is 420 cubits (220m). With a seked of seven, its original height was 210 cubits (110m). Its casing stones are laid in level courses with excellent fitting. Gaps between backing stones appear not as wide as in the Bent Pyramid, and are better filled with mortar and chip. Like the Bent Pyramid, it cannot be determined if it possesses any kind of nucleus structure.

Fig. 93 Casing stones, east side of the Red Pyramid.

It is natural to infer that whatever served as impetus for lowering the slope of the Bent Pyramid provided the incentive for starting this new building at the same, more conservative slope. But the compromise must have been a reluctant one. We can see in the next pyramids built, at Giza, adaptations accommodating steeper height-to-base proportions.

What purpose did the Red Pyramid serve? Perhaps Sneferu didn't want to be entombed in the damaged and "bent" pyramid to the south. He desired a more perfect structure for his eternal home. Nevertheless, he ordered the Bent Pyramid to be finished as a back-up tomb while work started on the Red Pyramid. He wasn't spreading his work force too thin. More than eighty percent of the volume of the Bent Pyramid is contained in the portion below the bend level.

The Red Pyramid's internal features are superbly executed but exhibit no architectural innovation.

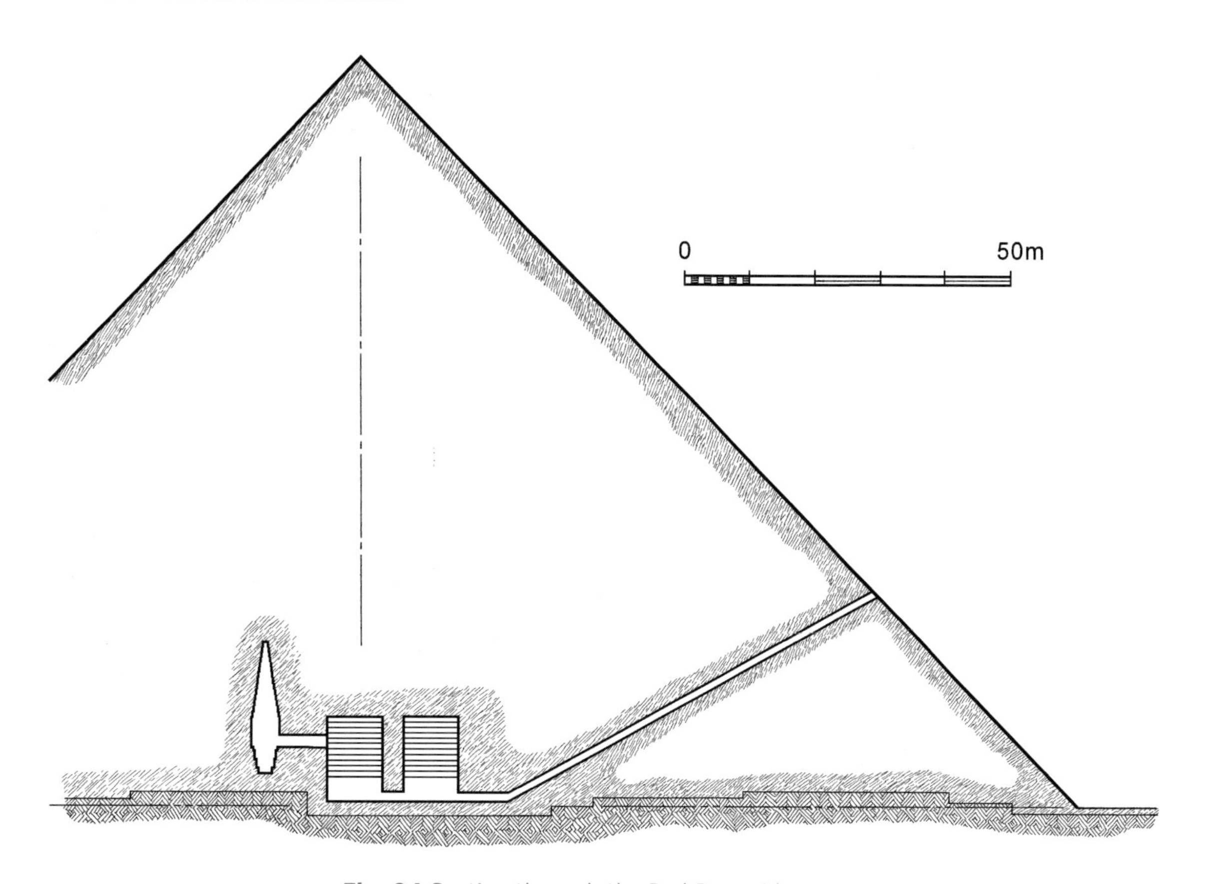

Fig. 94 Section through the Red Pyramid.

There are two puzzling things about the interior. The first is the lack of any sort of portcullis device. Perhaps the plan was to fill the entire entrance corridor with plug stones, making portcullis gates pointless. The second is the deep floor in the innermost room. From its entrance tunnel the floor of this room descends about three-and-a-half meters (10 feet), with walls corbelling inward almost in mirror image of the ceiling. Was this done to increase storage space in the room? If this room was the king's crypt, was his coffin to be placed in it? No one knows. Nor can we know if the king was buried here.

We do know that Sneferu's accomplishments were many. The first geometrically regular pyramids were built during his tenure. He produced the classic pyramid complex comprised of the main pyramid, mortuary temple, causeway, and valley temple. He added a substantial structure to one pyramid (at least) and built two more completely, thereby erecting the greatest volume of stone masonry by a single ruler in Egyptian history. Finally, he imbued in his son the drive that would culminate in the greatest pyramid of all.

7

GIZA

At places in this chapter I will use the designations of the Giza pyramids created by Reisner, where:

G1 = Khufu's pyramid, the Great Pyramid
G1a, G1b, G1c = the three small pyramids east of G1, north to south
G2 = Khafre's pyramid
G2a = the subsidiary pyramid south of G2
G3 = Menkaure's pyramid
G3a, G3b, G3c = the three small pyramids south of G3, west to east

KHUFU'S PYRAMID

Sneferu's son Khufu built the largest and most famous monument in history, the Great Pyramid. The architect is believed to be Khufu's nephew Prince Hemiunu (HEM-ū-new), son of Khufu's brother, Nefermaat. It is not just its size that has inspired ageless wonder. Its remarkable features, inside and out, make it the most studied, written about, and argued about building of all time.

The pyramid is enormous. Each side measures 440 cubits, or 230.38 meters (756 feet). Figure 95 shows the base area of the pyramid in comparison to an international football pitch (soccer field) of 68 x 105 meters, and at lower right, an American football field of 53 x 120 yards.

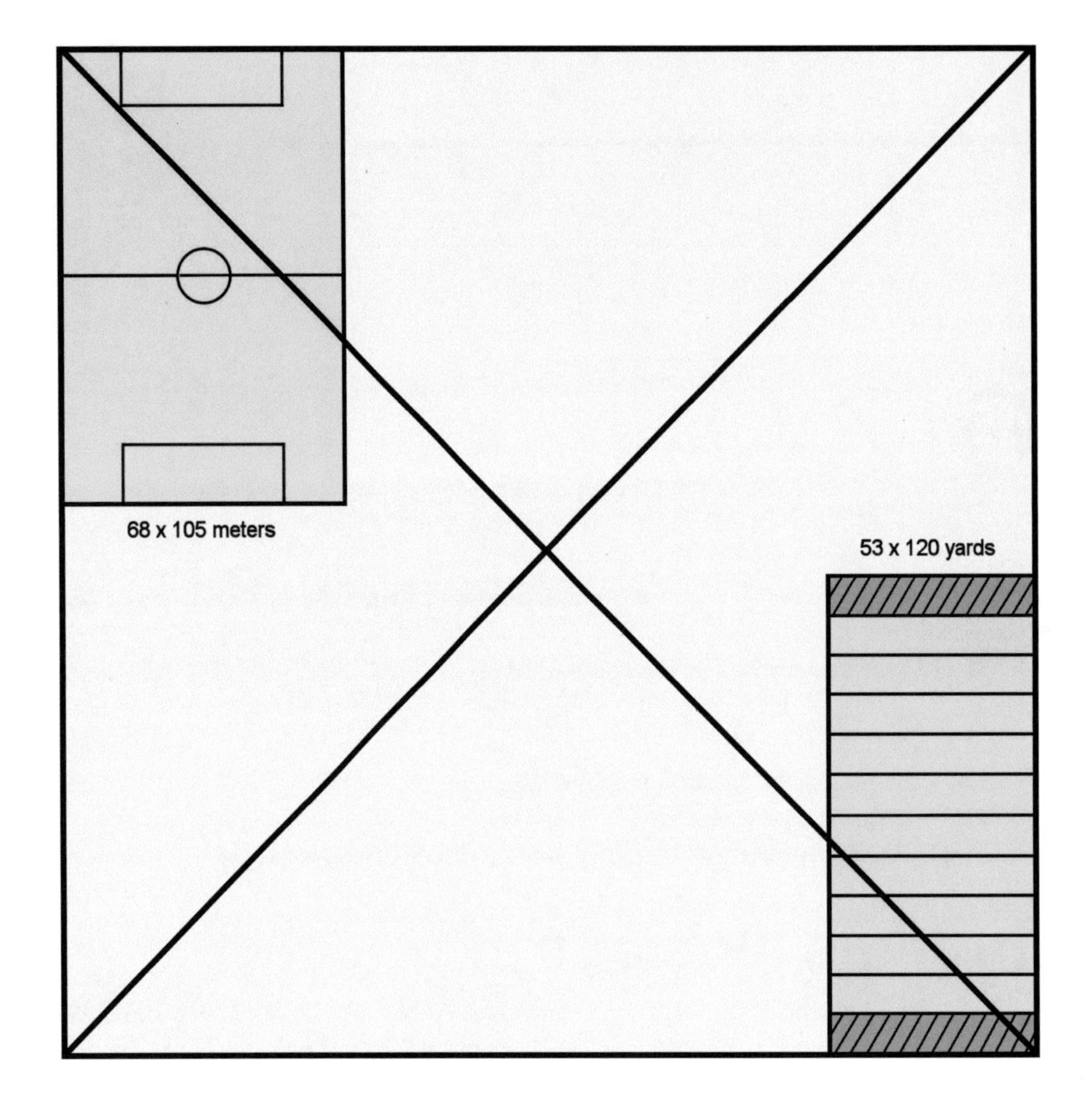

Fig. 95 Base area of the Great Pyramid.

To increase its majesty Hemiunu reverted to a seked of five hands two fingers, same as the ill-fated Meidum E3 addition. Its height therefore reached 280 cubits, or 146.6 meters (481 feet), and its volume nearly 2.6 million cubic meters. The area of its four faces was 21 acres. The pyramid originally contained about two million blocks with average weight of 2.6 tons.

Fig. 96 The Great Pyramid, from southwest.

The pyramid is founded on rock of the Giza plateau. In several places along its perimeter one can observe bedrock that has been cut to receive casing and backing stones. To support the first course of casing a pavement of fine white limestone was let into the rock all around the pyramid.

Fig. 97 Casing, backing, and pavement stones.

The pavement is about a half-meter thick. From the edge of the casing it extends outward thirty to forty-five centimeters (1-1.5 feet), and inward about two to three meters (6-10 feet). Though it varies considerably in thickness, its upper surface is remarkably level and flat. According to surveyor J. H. Cole (1925), not more than twenty-one millimeters (7/8ths of an inch) separates the highest and lowest points around the entire perimeter of 921 meters (3,023 feet).

The builders saved labor by incorporating a domed portion of the plateau into the body of the pyramid. In the "Grotto," described later, the knoll height is about six meters above zero level.

Today the pyramid is missing about nine meters of its original height. Its top surface, discounting the two partial courses in its center, is about 138 meters above zero level and is roughly ten meters square. With casing, this platform, the top of course 201, would have been about twelve meters square.

The casing blocks

Only a few casing stones remain. The best-preserved are in a row at the center of the northern base. The blocks from the middle to western end of this row have been reconstructed in the modern era.

The joints between casing and backing stones appear to be filled with white gypsum mortar, rendering them nearly invisible. We do not know if the mortar extends all the way across the joints. Nevertheless, the joints look to be no greater than a millimeter wide, an amazing degree of perfection considering the size of the blocks. For example, the casing block in the foreground of figure 98 has a mass of nearly 16,000 kilograms (17.6 tons). I will say more about casing stones in chapter nine when I address block handling and masonry techniques.

Fig. 98 Casing stones at north base of the Great Pyramid.

The backing stones behind the casing are important because they provide part of the foundation for the next course. Dietrich and Rosemarie Klemm say "there is no significant difference between the provenance of casing and backing material for the Khufu pyramid." Nevertheless, as I stated in chapter two, I found backing stones of local nummulitic limestone in many places on G1. The backing stones that run up the four corners of the pyramid do appear to be Tura stone, as are many backing stones on the lowest courses and the entire course now exposed at the top of the pyramid.

The entrance

The original entrance (**E** in Fig 101) to the pyramid is on the north face, a little over seven meters east of the north-south axis, and seventeen meters above zero level. Today many blocks that surrounded the entrance passageway are missing. Greek geographer Strabo (c. 64 BCE – 24 CE) described the entrance as follows: "High up, approximately midway between the sides, it has a moveable stone [door?], and when this is raised up there is a sloping passage to the vault." Strabo said nothing more about the passage, so we must wonder about the configuration of the movable stone and if he really witnessed its operation. He also said of the two largest pyramids that "their height is a little greater than the length of each of their sides," a ratio which is significantly backwards.

Fig. 99 Original entrance of G1 (upper left) and present tourist entrance (lower right).

In 1638, British mathematician/ astronomer John Greaves (1602-1652) traveled to Egypt to explore the pyramids. In his book *Pyramidographia* (1646), he includes a sketch of the entrance looking pretty much the way it looks today, so if the pyramid originally had a moveable door, the door and its surrounding masonry had disappeared by that time.

The tourist entrance is a rough-cut tunnel (**RT**) almost certainly made by the first violators of the pyramid. Why? Because the robbers made their tunnel horizontal, on the vertical axis of the pyramid, and level with the last plug block in the ascending passage. At its southern extent they turned the tunnel east to intersect the ascending passage. It looks like they knew where they wanted to go. As I mentioned in chapter two, Egyptologists believe all the pyramids were plundered during the First Intermediate Period, and Khufu's was no exception. After all, the pyramids were like giant billboards saying "the treasure is here."

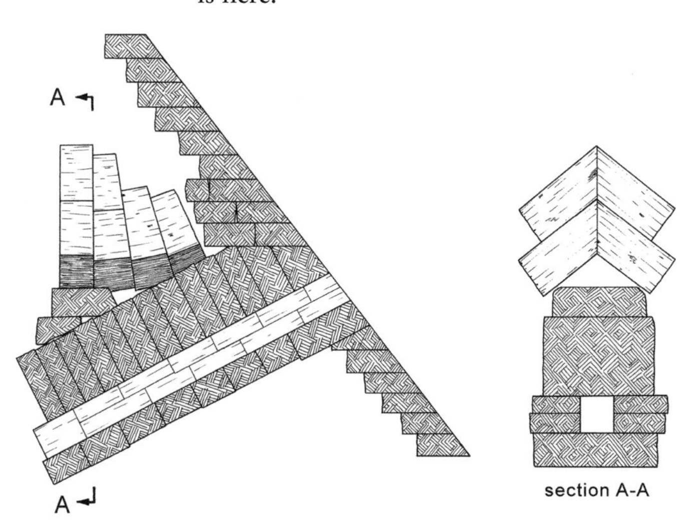

Fig. 100 System of protecting blocks over the entrance of Khufu's pyramid.

The builders took exceeding care to protect the entrance from crushing forces (if that was their purpose). They placed a series of gabled beams to direct the pressure of the overlying mantle around the corridor to the masonry below. Only the innermost set of beams is still in place. These stones have vertical faces, but we can detect from the foundations of their missing outer neighbors that three more sets of beams progressively leaned inward, and had their gable angles decreased in transition to the casing slope. Behind and under the gabled beams the corridor ceiling is formed by huge architraves. These blocks alone seem more than adequate to guarantee integrity of the passageway.

The entrance treatment, like the crypt ceiling described later, looks unnecessarily complex and overly cautious. Why were the builders so conservative? Was the reason other than guaranteed safety? Could it have been the desire to impress (the king, the gods) by their *tour de force*? This is idle speculation. We will never know.

Descending Passage and Subterranean Chamber

The Descending Passage (**DP**) slopes down from the entrance for 105 meters (344.5 feet). Petrie determined (1885, 19) that its mean angle is 26° 31' 23", a gradient of one rise to two run. The axis of this "polar passage," if extended to space, is about three-and-a-half degrees away from the celestial North Pole. This led British astronomer Richard Proctor (1837-1888) to his theory that the Great Pyramid was an observatory. I think it is probable, though, that pyramid architects designed sloping corridors for a more pragmatic need. I'll reveal it in chapter nine.

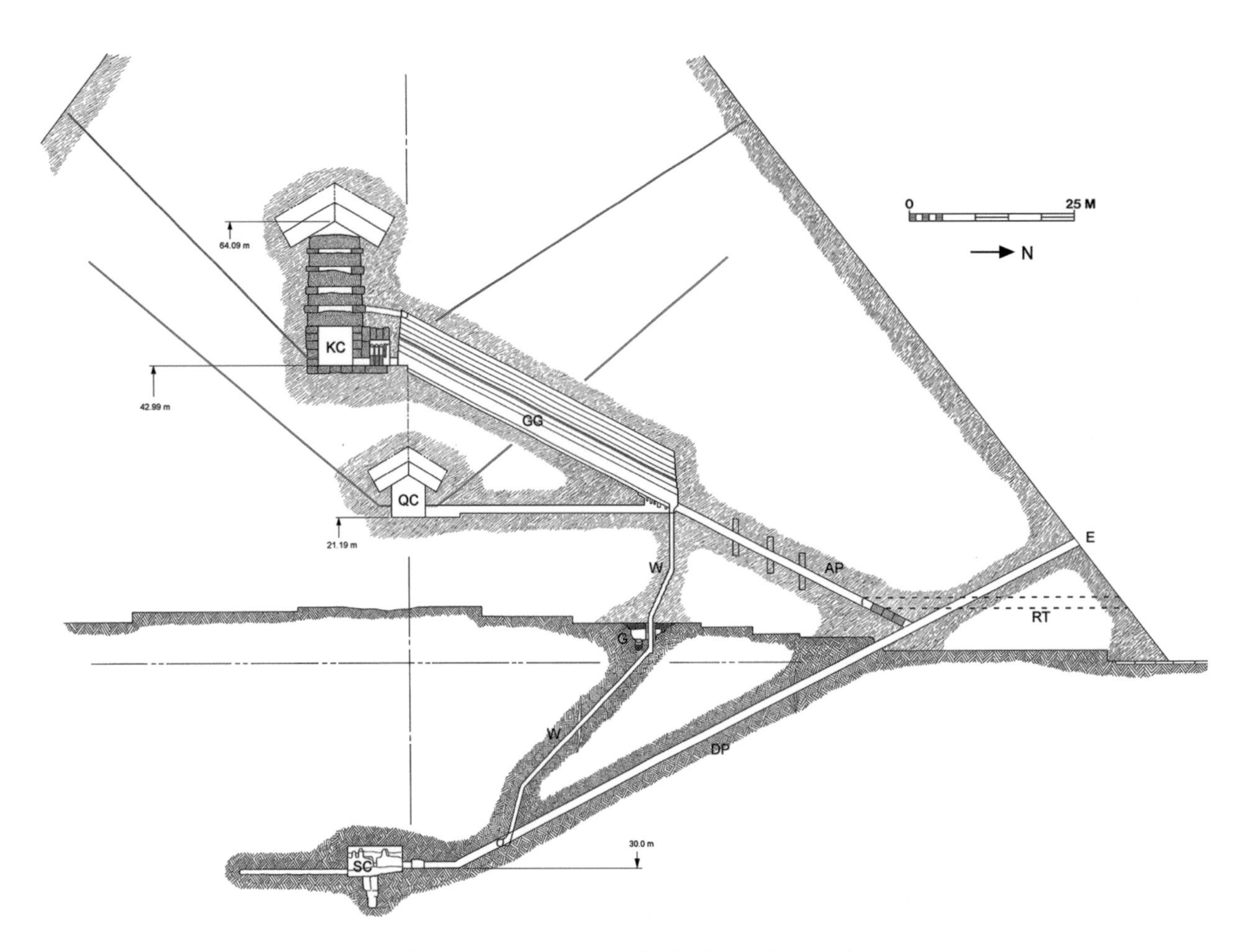

Fig. 101 Section through the Great Pyramid.
Redrawn from Maragioglio & Rinaldi, vol. IV (1965, Tav. 3).

Workmen sank the Descending Passage deep into bedrock, past the vertical axis of the pyramid. They started hollowing out the Subterranean Chamber (**SC**).

Fig. 102 The unfinished Subterranean Chamber.

This room was never completed. Its purpose is unclear. Petrie thought the unfinished room was intended to keep tomb robbers from discovering the real crypt above. Some have proposed that this room was to be the crypt in the first of three plans for the building. They think the second plan was to make the Queen's Chamber (**QC**) Khufu's burial vault before settling on the King's Chamber (**KC**) for that purpose.

The change-of-plan theory does not have evidential or logical support. Maragioglio & Rinaldi observed that the large blocks made with superb joinery at the junction of the Ascending and Descending Passages were necessarily early components of the pyramid. I could say the same for the junction where the Ascending Passage meets the passage to the Queen's Chamber at the base of the Grand Gallery. Those aspects, coupled with the meticulous planning, design, and construction evident in the upper chambers, persuade me that all internal features of Khufu's pyramid were planned from the start.

Perhaps the Subterranean Chamber was a provisional crypt in case Khufu died before the upper apartments were completed. We could speculate forever, or simply say that for a reason we don't understand, the room was either abandoned or meant to remain unfinished.

The Subterranean Chamber shows how the Egyptians excavated a room in bedrock. Their method was similar to that used in the underground quarries across the Nile. They first carved a horizontal slot delineating the ceiling, then trenched around its perimeter to create the walls of the room. They made vertical trenches through the material in the center of the room, breaking off large chunks of the resulting islands as they went along.

In at least six places on the ceiling of this room there appear shallow incisions that Mr. A. Pochan, in *The Mysteries of the Great Pyramids* (1971), called "mysterious letters" because they look like the letter M or W. I believe they are places where wooden beams were wedged to provide temporary shoring of the ceiling during excavation of the room.

Fig. 103 Shoring location in the ceiling of the Subterranean Chamber.

Though it seems improbable that solid rock would collapse, there was no guarantee that a large flake wouldn't fall on the workmen. Early on, the excavators were especially vulnerable to such an event. They were working in a vertically confined space probably less than a meter high, but of a breadth that eventually included the entire ceiling.

Fig. 104 Excavating the Subterranean Chamber.

The workmen forced wooden props into the ceiling cuts as they worked their way across the room. Each wedging incision was made with pockets for two props. The floor was cut down alternately between the two ceiling support points so at least one prop was in position at any time.

Hollowing out the Subterranean Chamber must have been a slow, tedious process. The laborers had to compete with their oil lamps for precious oxygen in the confined space. The heat and stone dust must have made an unpleasant job almost unbearable. Short shifts must have been the rule. Air circulation may have been improved when the descending corridor became connected with another shaft called the "Well."

The Well and Grotto

The Well (**W**) is a tunnel just large enough to admit an adult person. It connects the lower end of the Grand Gallery with the west side of the Descending Passage, 98 meters (321.5 feet) from the original face of the pyramid. Its total length is 58.4 meters (191.6 feet) (Vyse). Except for short east-west portions at both ends, it is built on a north-south azimuth parallel to, but 2.3 meters (7.5 feet) west of, the plane of the other corridors of the pyramid.

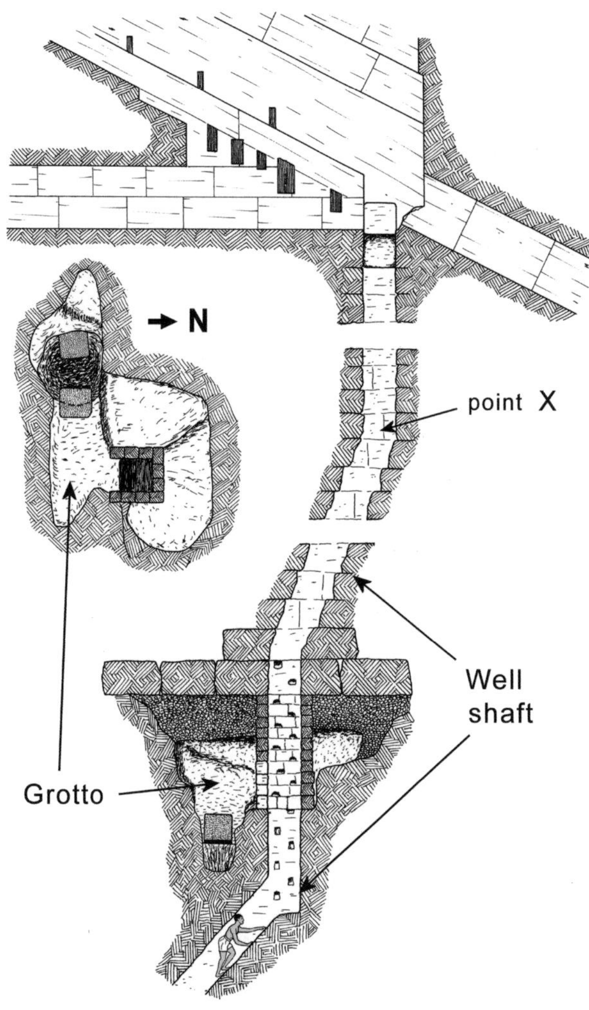

Fig. 105 The Well and Grotto in Khufu's pyramid.

The upper section of the Well shaft passes through the core masonry of the pyramid. The lower portion is cut through bedrock. At ground level, where the sections meet, a pocket in the rock, filled with a conglomerate of flint pebbles and sand, is traversed by a tunnel lined with small blocks of Tura limestone. At an unknown time someone made a doorway in the southern side of the masonry and hollowed out a portion of the filling, creating the Grotto (**G**).

Fig. 106 The Grotto, looking east toward the opening to the Well shaft. The granite block in right foreground beneath the author is about 0.5 m thick, and might be a piece of one of the sliding gates from the portcullis room described later.

Fig. 107 The Grotto, looking west from the entrance. The holed granite block in this photo (lower left) could be another piece of a destroyed portcullis gate.

Greaves mentioned the Well in *Pyramidographia.* He supposed that the Well might have been "the passage to the secret vaults hewn out of the rock upon which this pyramid was erected." He limited his exploration to dropping a line down the Well from the Grand Gallery. Finding its depth to be twenty feet, he said that beyond that point it was "choked up with rubble."

Over the years several attempts were made to find the bottom of the Well. In 1700, Benoit de Maillet, French Consul General in Egypt, climbed down past the Grotto to find a rubble obstruction 43 meters (141 feet) below the Grand Gallery entrance. This obstruction, it now appears, was largely the conglomerate removed in hollowing out the Grotto. Sixty-four years later, English traveler Nathaniel Davison reached the same point and, like Maillet, could go no farther. During Napoleon's occupation of Egypt in 1799-1801, Colonel Jean-Marie-Joseph Coutelle (1748-1835) was able to clear several meters of the shaft filling before the near impossible work forced his men to yield.

Success finally came in 1817. Giovanni Caviglia (1770-1845), a retired sea captain, continued the clearance of the tunnel before his men too were forced to give up, probably within a few meters of the Well bottom. Then, setting his workers to clearing the rubble in the Descending Passage, he exposed the lower terminus of the Well. He removed the remaining debris that blocked the tunnel, at last providing the unobstructed passage we have today.

On the history and purpose of the Well, theories abound in sublime profusion. As many are fanciful, I will not cover them all. The only points on which most writers seem to agree are that the Well had a purpose and that part, or all, of the shaft was not in the builders' original plan.

Borchardt thought all the upper apartments resulted from changes in the first plan, which was to bury Khufu in the Subterranean Chamber. Petrie, later supported by Maragioglio & Rinaldi, argued that only the Well was not in the original plan. In *The Pyramids and Temples of Gizeh* (1885) Petrie wrote:

> The shaft, or "Well", leading from the N. end of the gallery down to the subterranean parts, was either not contemplated at first, or else forgotten in the course of building; the proof of this is that it has been cut through the masonry after the courses were completed. On examining the shaft, it is found to be irregularly tortuous through the masonry, and without any arrangement of the blocks to suit it; while in more than one place a corner of a block may be seen left in the irregular curved side of the shaft, all the rest of the block having disappeared in cutting the shaft. This is a conclusive point, since it would never have been so built at first. A similar feature is at the mouth of the passage, in the gallery. Here the sides of the mouth are very well cut, quite as good work as the dressing of the gallery walls; but on the S. side there is a vertical joint in the gallery side, only 5.3 inches from the mouth. Now, great care is always taken in the Pyramid to put large stones at a corner, and it is quite inconceivable that a Pyramid builder would put a mere slip 5.3 inches thick beside the opening to a passage. It is clear, then, that the whole of this shaft is an additional feature of the first plan.

As I mentioned earlier, Maragioglio & Rinaldi adduced as proof that all the major rooms and corridors in the pyramid *except for the Well* were in the original plan, the fact that the corridor junctions are specially reinforced with unusually large blocks.

But why should the Well be an exception? I believe that the entire extent of the Well was planned from the beginning. My best guess as to its history and purpose follows:

The builders began excavating the Descending Passage shortly after the corners of the Pyramid were fixed. At the same time they commenced digging the shaft downward from the conglomerate-filled depression that would later form the Grotto. Why was the shaft started here? Perhaps because the plan said so. The masons lined the portion of the shaft that traversed the gravel pocket with walls of small blocks of Tura limestone bonded with mortar. They commenced digging the shaft down through the rock. The conglomerate filling in the Grotto is tightly packed around the masonry shaft lining, an indication that the blocks were placed first, followed by the filling around them. This would not be so if the blocks were placed against the fill, as the plan-change proponents believe.

The Well may have served several functions, but only one can I state with conviction: For those who pushed the plug stones (described in next section) down the Ascending Passage it was the only way out.

Another use may have been to provide fresh air for the workmen who were carving the Subterranean Chamber. In 1864, during the American Civil War, the Union army besieged the fortified town of Petersburg, Virginia. To bypass the fortifications, engineers drove a 511-foot tunnel to plant explosives under the Confederate army position. At a certain point in the tunnel the sappers built a small fire at the base of a vertical shaft that led to the surface above. The draft from the fire did a good job of drawing fresh air from the tunnel entrance. This same principle may have been used in construction of the Subterranean Chamber. Looking down from the Grotto, there is a small landing. A fire built there would have drawn air down the Descending passage and up through the lower portion of the shaft. The walls of the shaft at this point appear blackened by soot, so this idea is not crazy.

Still another purpose for the Well was suggested by Lehner: The shaft served as a measuring reference to help guide construction of the upper chambers of the pyramid. In *The Pyramid Tomb of Hetep-heres and the Satellite Pyramid of Khufu* (1985), Lehner proposed that "these shafts [vertical segments of the Well] may have served as a vertical control for the establishment of junctures between passages. The shaft offered a place for a plumb line and a determination of height – a datum which rose course by course with the pyramid body."

It came time to carry the Well upward. Limestone slabs were carefully placed to cover the gravel pit without crushing the small-block shaft lining. When I examined the Well in 1987 the segment above the Grotto did not appear "irregularly tortuous," as Petrie said, but is a straight shaft with protruding block corners. I think it was built this way to make it easier to traverse. It may not be clear in figure 108, but the large blocks in the dark area into which the rope disappears, those directly above the Grotto-traversing small-block masonry, are carefully mortared together. To me the passage looks *constructed,* not later-carved.

Fig. 108 Looking up the Well shaft from the Grotto. North is down.

The Well was inclined to the north as it rose until it reached a level I call point X in figure 105. Because point X was on the proper latitude for intersection with the north end of the Grand Gallery, the shaft was turned vertically upward here. Just below the lower end of the Grand Gallery the builders directed the shaft ninety degrees to the east, then upward, in line with the bench that runs along the base of the west side of the Gallery.

The masons didn't immediately connect the Well with the Gallery. Instead they covered the mouth of the shaft with an unobtrusive block that became part of the western bench. Why was this done? Two thoughts: The first, and least likely, is that Khufu was kept ignorant of the shaft that would allow escape of the workers who would seal the tomb. But considering the thousands of people working on the pyramid, a secret like that would seem hard to keep. It's more likely that Khufu condoned the Well because it served as a construction aid, and because its ultimate use assured proper sealing of the Ascending Passage. In the period between constructing the inner apartments and Khufu's death the interior was undoubtedly visited many times by the king and confidants. As there was no need to draw attention to the Well, its upper terminus was temporarily hidden.

When it came time to seal the pyramid trusted workmen chiseled down through the blocking stone to finally join the Well with the Gallery. One can see that the blocking stone was removed by chiseling from above.

Fig. 109 Dirt-ball Bob after the Well/Grotto excursion.
Note the flashlight, pointing down, taped to my right leg. Returning to the swanky Mena House hotel I garnered a few looks while passing through the lobby.

The plug stones

To begin sealing the pyramid, workmen lowered the granite portcullis gates in the little antechamber before the crypt. They assembled a wooden bridge across the mouth of the corridor that leads to the Queen's Chamber. They pushed three granite blocks, previously stored on the Gallery floor, one-by-one across the wooden bridge and down the Ascending Passage, sealing each with plaster on its upper side. Petrie and Maragioglio & Rinaldi noted that the holes in the sides of the Gallery for the cross-beams which would have supported the wooden bridge are large enough that the bridge could have taken the weight of the plug stones, a counter to some who believe the plug stones were built in place.

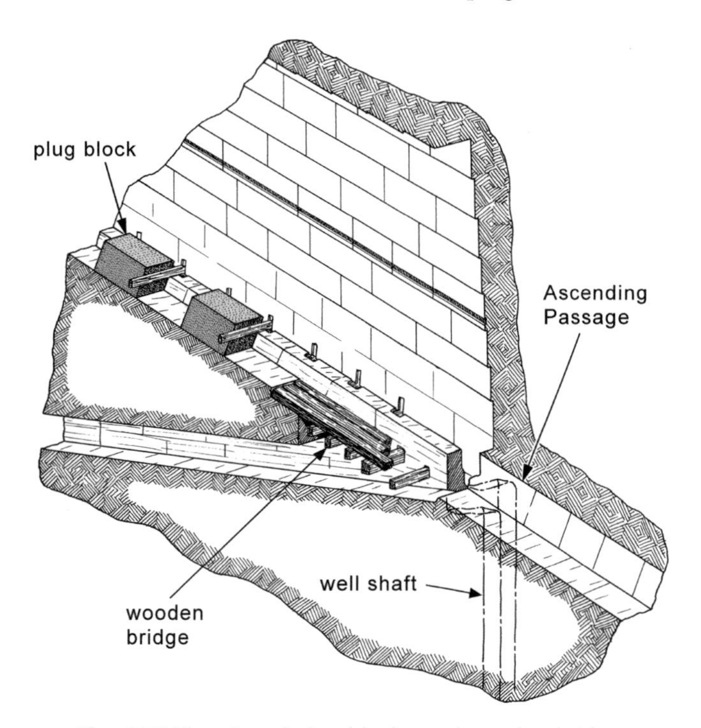

Fig. 110 The stored plug blocks and wooden bridge.

Might a limestone plug block have preceded the three granite blocks down the passageway, and therefore served to hide the presence of the ascending passage from tomb robbers? I doubt it. The block would not have been long enough to be stably retained in its final, trimmed form. It would have drooped, leaving a substantial opening at its upper edge, thus defeating the purpose of concealment.

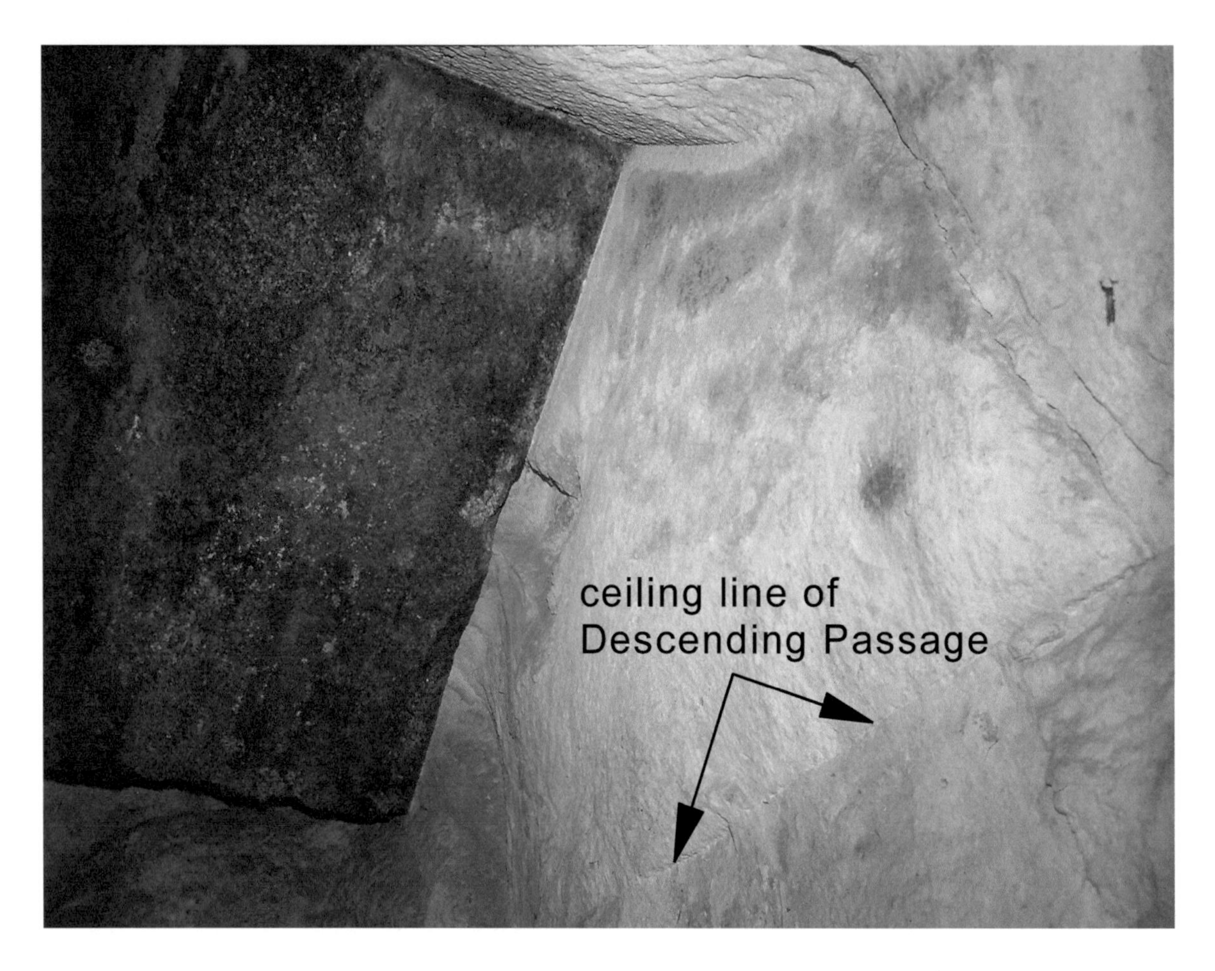

Fig. 111 On left, the face of the Granite plug block at beginning of the Ascending Passage. The arrows show the ceiling edge of the Descending Passage.

Tasks completed, the workmen departed the pyramid via the Well and Descending Passage.

No one knows if the Descending Passage was partly or totally plugged. Petrie thought the corridor was allowed to remain open to serve as a blind for robbers who would be purposely led to the disappointing Subterranean Chamber. If he was right, the intersections of the Descending Passage and Well with the Ascending passage would have necessarily been camouflaged somehow.

Back to the Well. Maragioglio & Rinaldi thought the Well was a "service shaft" not in the original plan of the pyramid. They said the pyramid had reached the level of point X when the builders decided to create the shaft. The workmen dug downward at a steep angle until they encountered the loose conglomerate that filled the Grotto. The diggers then altered the direction of the tunnel to vertical to facilitate construction of the small-block walls that retain the crumbly material around them.

The Italians said that the shaft was not an escape route for the sealing crew. They believed it was used for transit of workers and materials between the Subterranean Chamber and the upper apartments. They said (vol. IV, 140) "it is evident that passage (P) [the Well shaft] was not an escape way for the men who had to place the granite plugs because at the moment of the king's funeral it was already closed." Further, they thought the shaft was filled from bottom to top and plugged at both ends before the pyramid was sealed. They opine further that workmen operating from the outside (lower end) of the granite plugs allowed the string of stones to slide down the Ascending Passage. Progress of the blocks was controlled with wooden wedges, as proposed by Georges Goyon.

Besides my previously stated reasons why I think the builders planned the Well from the start, I have the following objections to the hypothesis of Maragioglio & Rinaldi:

1. The inclined portion of the Well shaft that descends from point X turns vertically down about a half meter *before* it enters the walled Grotto lining. If the builders hadn't planned the Well they would have had no reason to remember the exact level of the gravel pocket, so it is doubtful they would have anticipated its presence and turned the shaft vertically downward immediately before the gravel was met. Further, the masons who constructed the block lining through the gravel apparently were aware of the greater extent of the pocket toward the west and north. They buttressed these sides with thicker blocks, a circumstance also implying they knew the gravel pit configuration.
2. The Well was not a service shaft because there was nothing to service. It's hard enough to get one's body up and down the shaft without carrying anything. Thus, it's hard to imagine that materials were regularly taken through this tunnel.
3. There is no evidence that the Well was even partly filled by the ancient builders.
4. The sealing of an ascending passage without the help of workers behind the plug stones had been tried before, without success. In the subsidiary pyramid on the south side of the Bent Pyramid plug stones were released by a remote mechanism. No escape shaft is present. Though its passage slope of 31.5 degrees is considerably steeper than that found in G1, only the first of three stones slid down the corridor. This fact would not have been lost on the careful architects of the Great Pyramid. I doubt they would have accepted that risk.

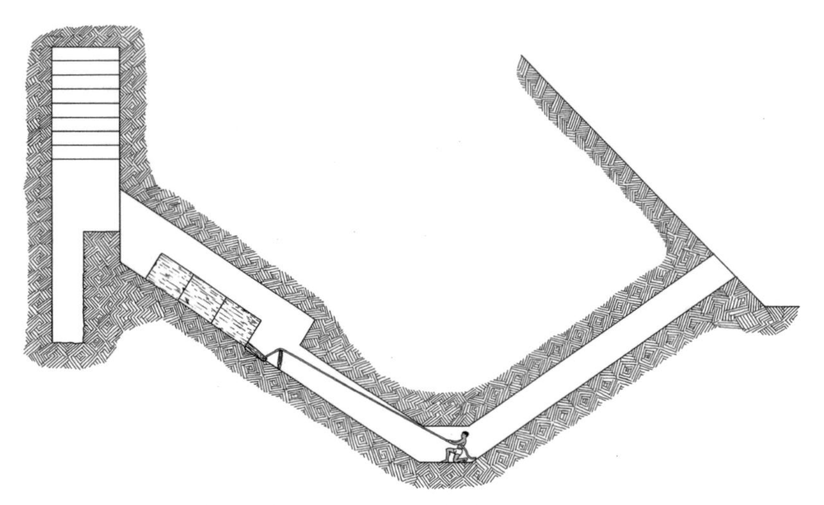

Fig. 112 Subsidiary pyramid of the Bent Pyramid.

The Ascending Passage and girdle blocks

The Ascending Passage (**AS**) rises at the same slope as the Descending Passage, two units of run for each unit of rise. The builders anchored this shaft to the core masonry with huge "girdle" blocks. These blocks prevent accumulation of compressive loads along the corridor, and thus reduce the chance of displacement or fracturing of blocks at the lower end of the passage.

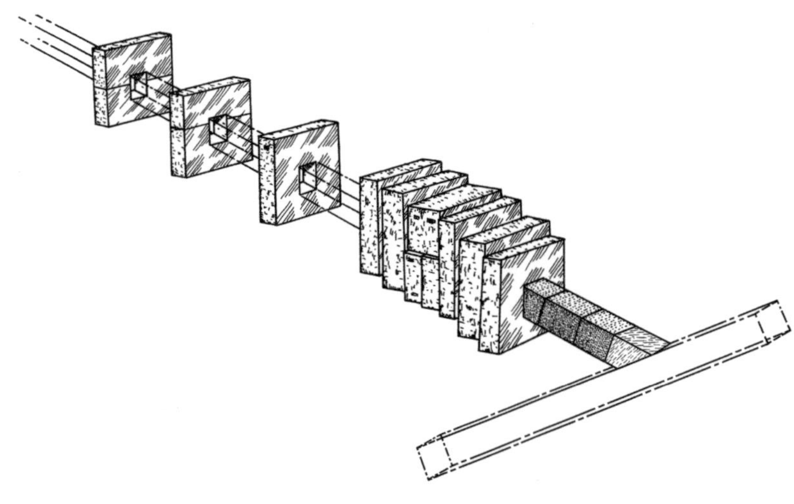

Fig. 113 The girdle blocks of the Ascending Passage.

While the girdle blocks at the lower end of the passage abut, those in the middle are spaced ten cubits apart. The spaced girdle stones led Borchardt to suggest that their positions marked the faces of an internal structure of large steps. Though I agree that a stepped nucleus is possible, perhaps probable, I doubt the girdles mark its faces. The internal passages and rooms would likely have been constructed in an area of sufficient width, called a "construction gap" by Arnold, that the step faces would not interfere with the work.

The Grand Gallery

The Grand Gallery (**GG**) is the largest room in any pyramid. Its ceiling rises 8.6 meters (28.2 feet) from the floor. Its slant length is 48 meters (157.5 feet). At floor level its width is two cubits, the same width as the Ascending Passage, but it has "benches" on either side of its full length. At the top of the benches the Gallery widens to four cubits (2.1 meters).

Fig. 114 North end of the Grand Gallery.

Why does the Gallery have such a high ceiling? A ceiling half as high would have allowed plenty of room for traversing the stored plug stones. The answer probably hinges in part on the width of the room. Possibly to allow space to maneuver plug blocks the Gallery had to be two cubits wider than the other corridors. But the masons deemed a width of four cubits too great to be spanned by simple beams. Thus, they reduced that width by successively stepping each of the seven wall courses inward by one hand. The ceiling width then matched the passage width between the shelves, two cubits.

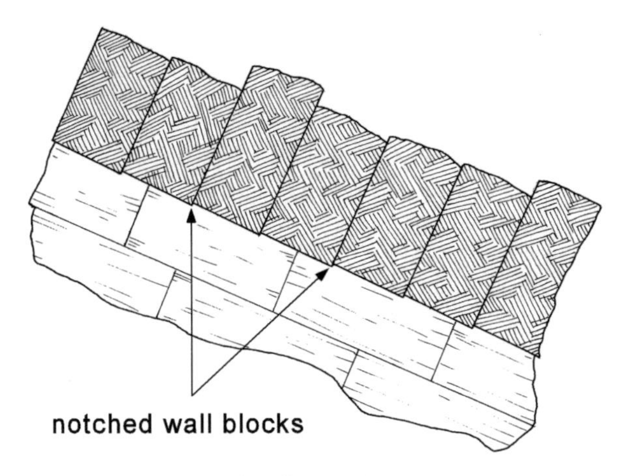

Fig. 115 Grand Gallery ceiling construction.

The transverse ceiling beams of the Gallery find notches in the tops of the walls. Like the girdle stones described before, the wall notches transfer loads locally and so prevent accumulation of stress at the lower end of the Gallery.

Spaced along the bench tops in the Gallery are odd-shaped cuts (shown in figure 110) of unknown purpose. At each cut a short vertical groove appears in the wall. Some have suggested that the notches secured the plug stones by means of wooden crossbeams before the builders were ready to seal the Ascending Passage. Maragioglio & Rinaldi argue that this purpose was unlikely because the notches continue in the area of the wooden bridge – not a good place to store plug blocks. They suggested that the notches may have held a wooden framework that supported some sort of decorative canopy. That may be so, but a wooden framework would have had another use during construction of the Gallery. It may have served to prop the walls apart before the ceiling beams were placed. On opposite walls of the third offset course, continuous grooves run the full length of the Gallery. Wooden beams placed in those grooves, braced by other beams crossing the room, would have kept the walls from collapsing inward before the ceiling beams were placed.

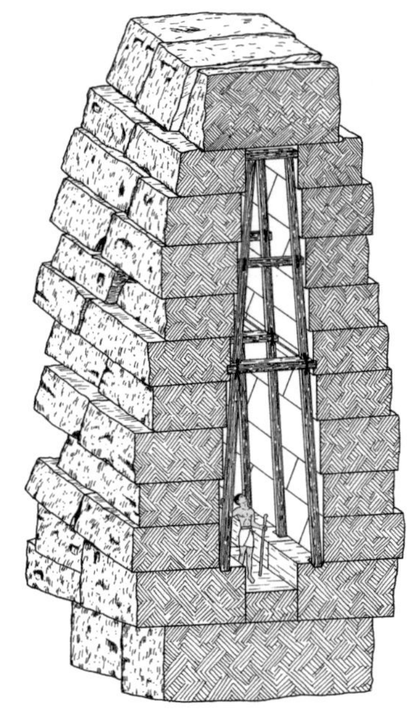

Fig. 116 Grand Gallery construction.

Another possible use of the continuous grooves is they may have contained planks that formed the floor of a storage area for funerary goods. All of these ideas on the purpose of the Gallery are, of course, speculative.

The Queen's Chamber

At the lower end of the Grand Gallery a horizontal passage leads south to the Queen's Chamber (**QC**), the name a modern misnomer that has, nevertheless, stuck. Egyptologists believe that no queens were interred in the large pyramids. For that matter, it has not been proved that any kings were either. Petrie proposed that this room was a *serdab*, a dwelling for the king's *ka* statue. In the east wall is a large, well made corbelled recess that may have contained this statue. In an age past, folks probably thinking the niche was a doorway to hidden treasure, made a deep tunnel in its sunken face. Theirs was likely a disappointing exercise.

Fig. 117 Niche in east wall of the Queen's Chamber.

The peaked ceiling of the Queen's Chamber is formed of six rows of butting limestone beams. The lower ends of these beams thrust upon foundation pads (the *springs* of the arch) located well beyond the walls of the room. In other words, the walls do not support the ceiling. The top courses, at least, were inserted after the ceiling was made.

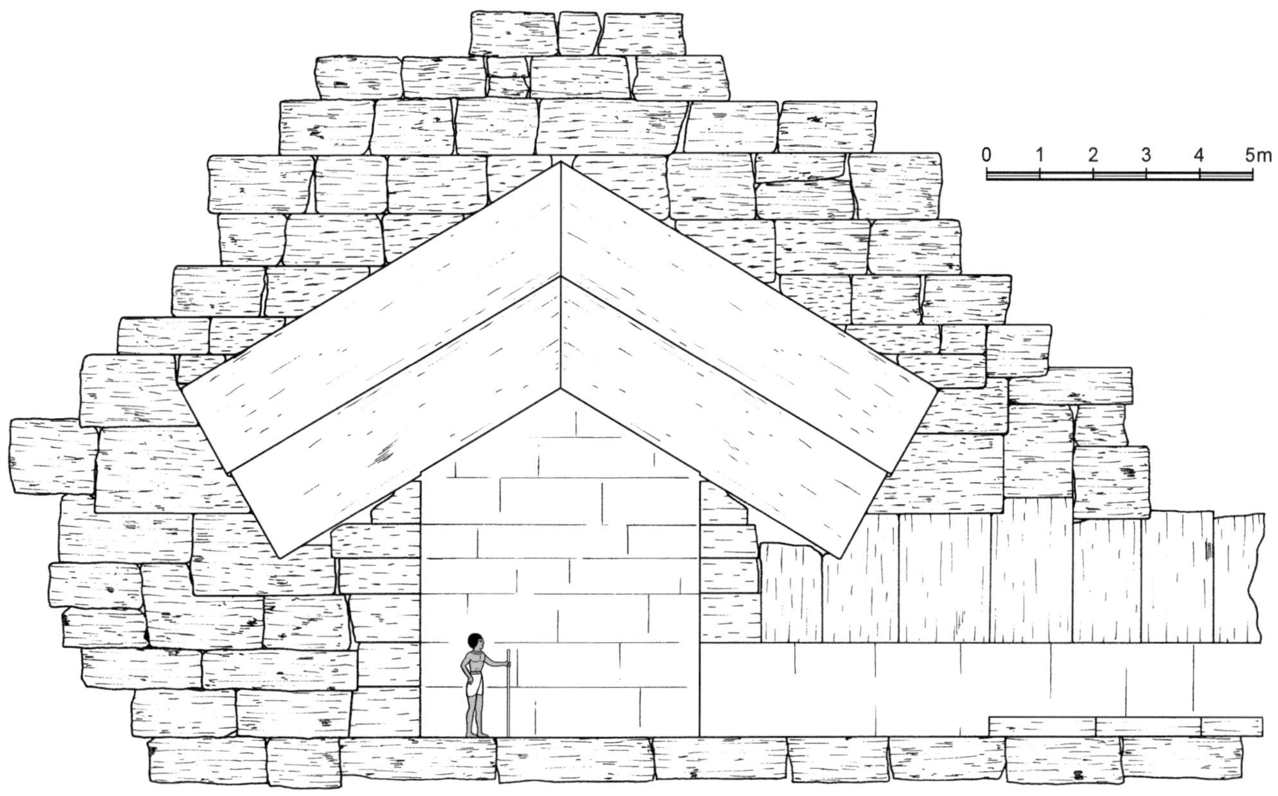

Fig. 118 The Queen's Chamber.

The masonry arch

In the ceiling of the Queen's Chamber we should recognize a simple but true masonry arch containing but two voussoirs per row. It's likely that another arch lies above the one we see. Most burial chamber roofs of this type that have been uncovered in pyramids built after Dynasty 4 possesses at least one additional ceiling, and several have a total of three.

Though many would credit the Etruscans and Romans with perfecting the masonry arch as depicted in figure 119, as far as I can determine, its earliest use was in this building. The invention is important. Stone is a superb building material, but it does not possess high tensile strength. *Tension* is the force trying to stretch a material, or pull it apart.

Tensile stress is always present in the lower part of a horizontal beam, even if it supports no load, because of its own weight.

An attribute of stone in building construction is its high strength in *compression*, a pushing force. The arch takes advantage of this quality by squeezing the stone so hard on its ends that the tensile stress at its lower surface is greatly reduced or eliminated.

Fig. 119 The masonry arch (voussiors and central keystone shaded).

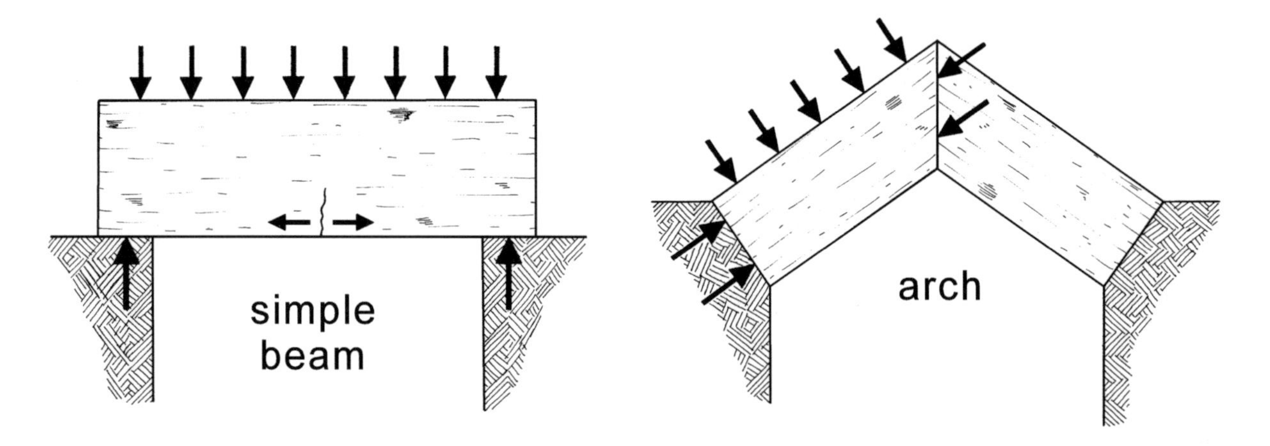

Fig. 120 Ceiling forces on the simple beam versus the arch.

The superiority of the arch is dramatically proved by comparing the ceilings of the Queen's and King's Chambers. The arched limestone ceiling of the Queen's Chamber is in good condition. Conversely, all of the simply-supported granite beams that span the King's Chamber are cracked through. Only tremendous compressive forces on their ends keep these great beams from falling.

The portcullis room

At the upper end of the Grand Gallery a short horizontal corridor passes through a little portcullis room that guarded the King's Chamber. The portcullis room, and everything onward, is built of Aswan granite. In the portcullis room, three granite gates, now missing, operated in vertical tracks in the walls.

Fig. 121 The portcullis room, looking up and northwest.

Fig. 122 The portcullis room, looking up toward southeast.

It appears that the stone gates were lowered by ropes that were wrapped around wooden logs fixed just above. Sockets for the logs appear in the west wall of the chamber. A flat shelf upon which the logs could rest appears on the east wall. In the south wall are four grooves that prevented pinching of the ropes that were used for lowering the southernmost gate. Also, the south wall has been chipped away in the center of its lower edge where the short corridor to the King's Chamber begins. Figure 123 is a diagram of the portcullis room as it appears today.

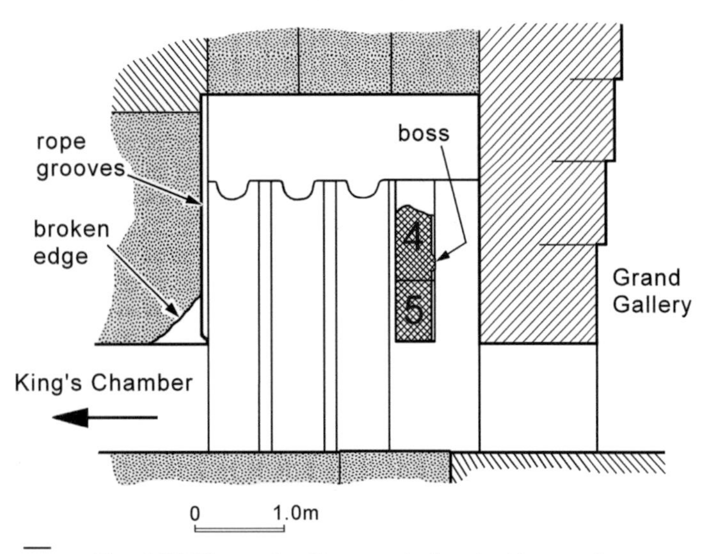

Fig. 123 The portcullis room today, looking west.

There are two slabs of granite still in place at the north end of the chamber. I have numbered them 4 and 5 in figure 123. The bottom slab sits on recesses in the walls, so it was not meant to be dropped. I will get to the top slab in a moment, but first, what was the purpose of these two slabs? I think they were to prevent easy access of tomb robbers to the innermost (southern) of the three portcullis gates.

If the robbers could have simply climbed over slabs 4 and 5 then those slabs would have offered no protection at all. Their presence would be pointless. Alternately, if slab 4 extended all the way to the ceiling, this problem would disappear, but another would arise: there would have been no way to place ropes around the logs in the process, years later, of lowering the three main gates.

How, then, can we find a useful purpose for these two slabs?

Notice that the upper slab has a broken top surface and a protruding boss on its north side.

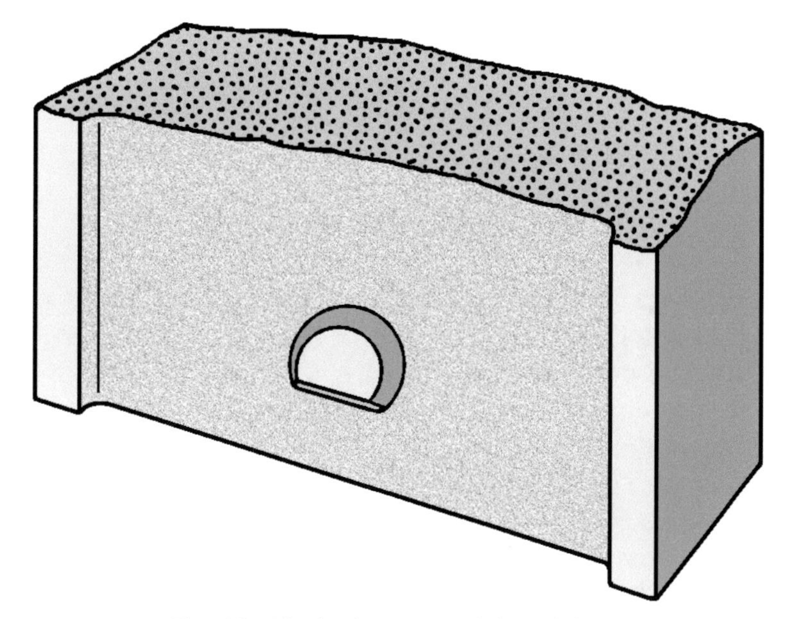

Fig. 124 The broken upper slab, with boss.

Ignoring the boss for a moment, what is the significance of the broken upper portion of slab 4? What is missing and why was it removed? I believe this slab had a protruding feature that denied robbers immediate access to gate 3 by blocking their path across the top of the portcullis chamber. A possible shape of the protrusion is shown in figure 125, the probable aspect of the portcullis room during the time the upper parts of the pyramid were being constructed and before the King's Chamber was sealed.

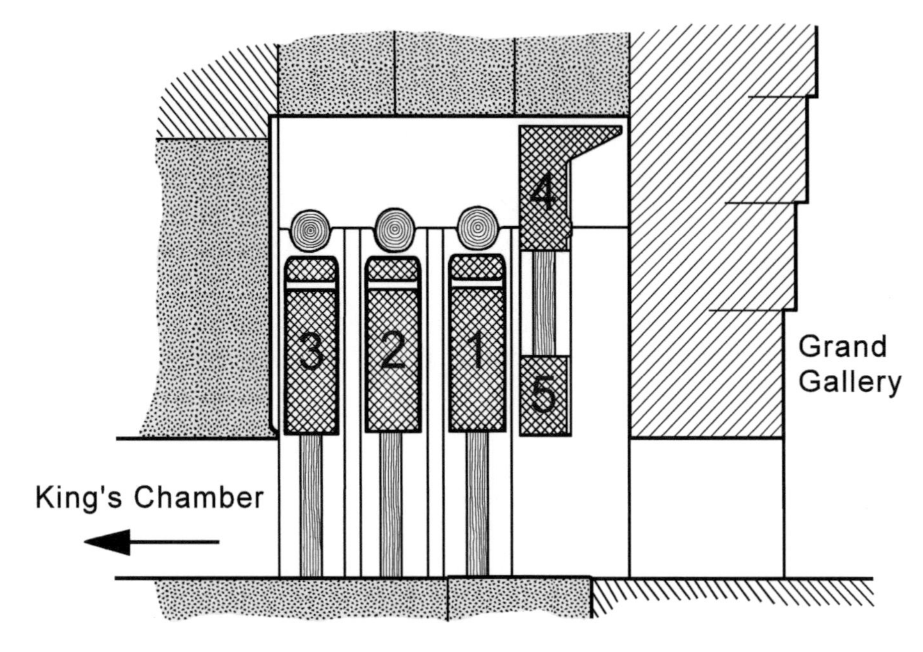

Fig. 125 The portcullis room before sealing.

Blocks 1 thru 4 were propped with wooden beams on both ends to prevent accidental release during the several years while the pyramid was open. The props under the main gates, 1 through 3, were envisioned by Maragioglio & Rinaldi. The props under block 4, and what block 4 originally looked like, are my suggestion. The logs shown above the gates are also hypothetical.

When it came time to seal the pyramid the builders lever-jacked gate 1 up enough to remove the two wooden props, and lowered it to two shorter ones. Repeating this procedure, they lowered gate 1 to about halfway down. Now a workman could craw through the opening between slabs 4 and 5 and place ropes which would be used to finally lower the three portcullis gates. The rope arrangement is shown in figure 126. The logs above the gates were round not because they functioned as rollers, but because this shape prevented the braking ropes that would slide around them from being unduly stressed. For clarity I have shown the rope arrangement only for gate 2.

Tension was applied to the ropes by workmen in the Grand Gallery. After all three gates were supported by ropes, workmen could crawl past gate 1, which was still held halfway up by its wooden props, and lever-jack gates 3, 2, and 1 up enough to remove the wooden props.

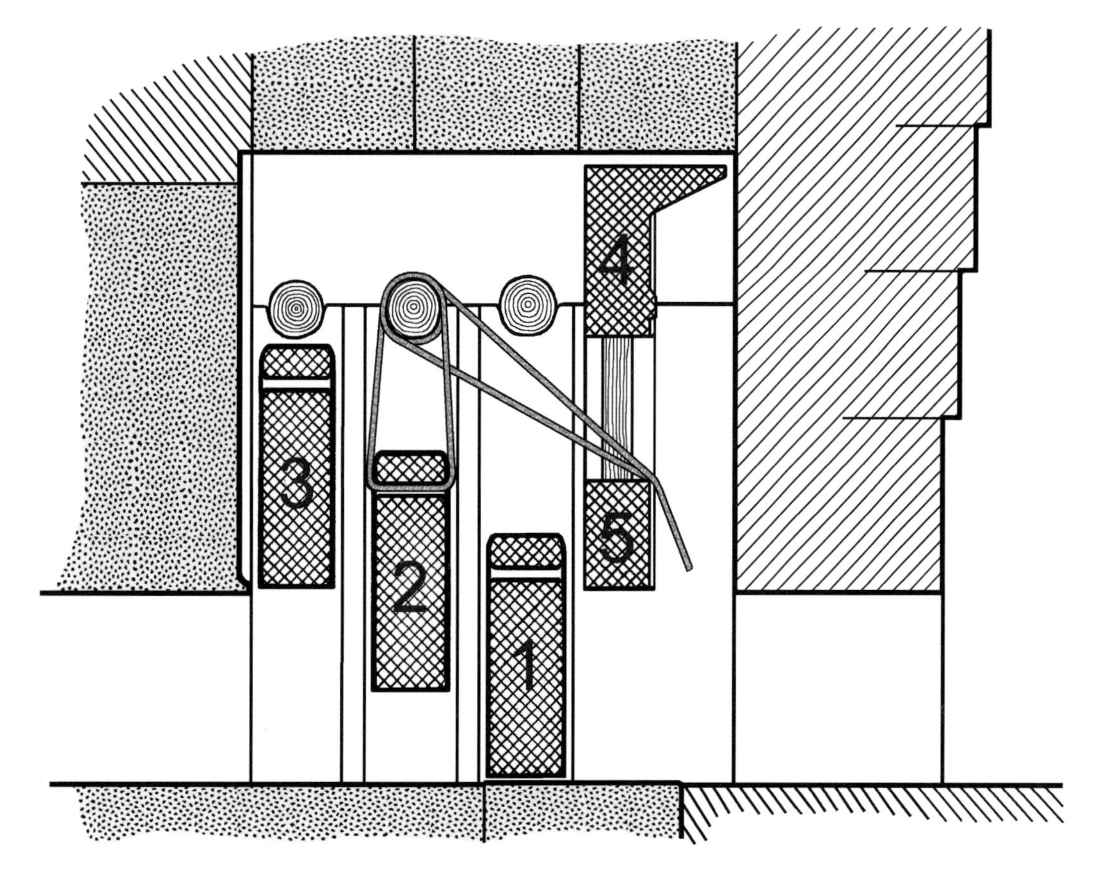

Fig. 126 The arrangement of ropes used to lower the portcullis gates.

With props out of the way, gate 1 was lowered to the floor. Note that the braking ropes were disposed such that they did not have to slide through the holes in the gates. After gate 1 was lowered there was room to observe the lowering of gate 2, then gate 3, the order being the opposite of what one might guess by intuition. All ropes were removed. The logs above the gates were cut to shorter lengths and removed through the opening between slabs 4 and 5.

It was time to lower slab 4. It was lever-jacked up enough to remove its two wooden props, then lowered to two shorter props. A series of ever-shorter jacking posts and props could then be used to repeat the same procedure I previously described for lowering gate 1 halfway down. The jacking post and lever by which it was pushed upward and then let down are shown in figure 127.

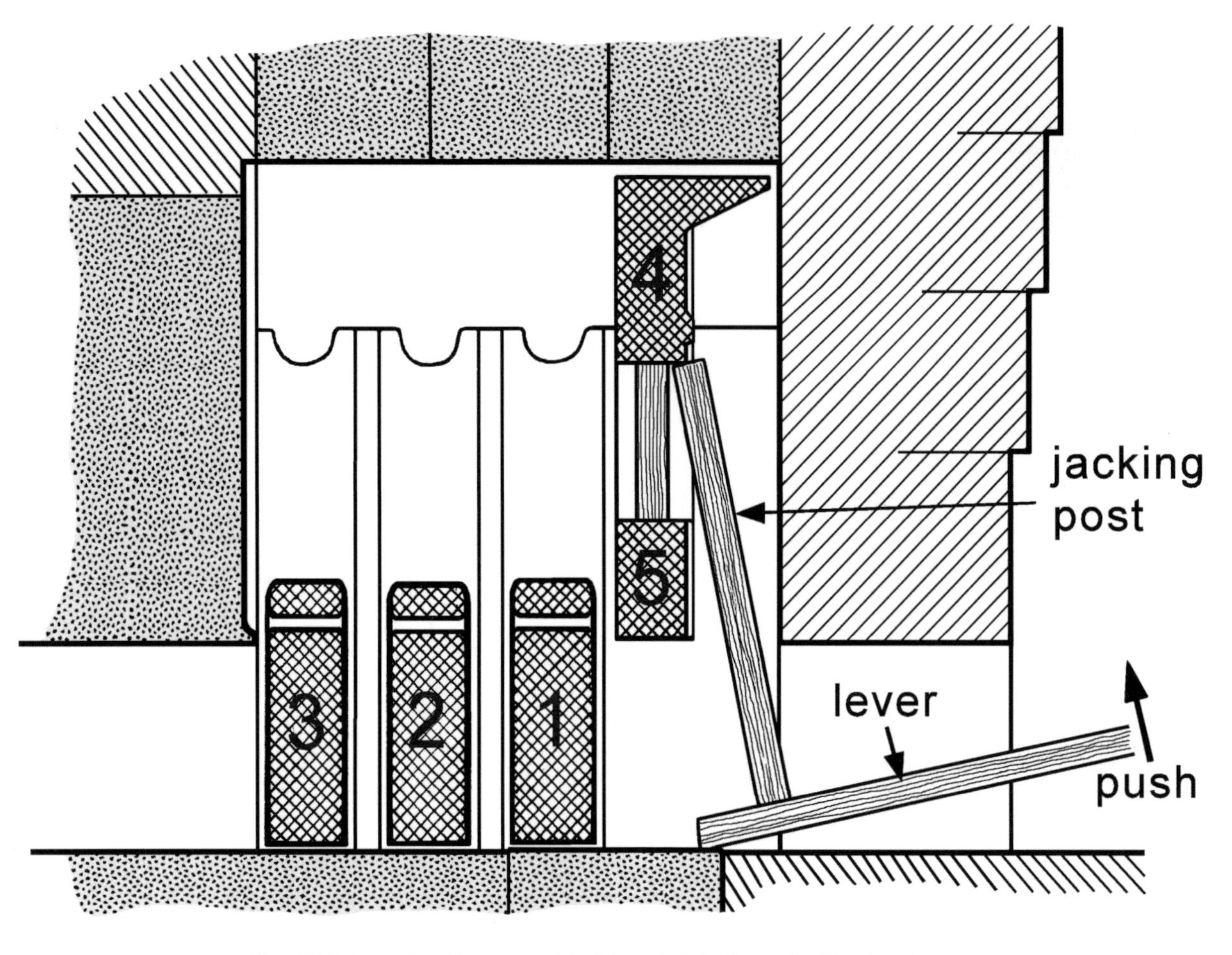

Fig. 127 Lowering the upper blocking slab by lever-jacking in stages.

When slab 4 got to within say 25 cm (10 inches) of slab 5 the need for the previously-mentioned boss arose. The jacking post had so little purchase on slab 4 that it might fracture, leaving bits of wood between the two slabs that would be difficult to remove. The overhanging top portion of slab 4 was not a good place to put the jacking post because a high force there would put a torque on the slab that could break off its rear guide rails. The boss on slab 4 became the safest place to put the jacking post for the last bit of lowering the slab.

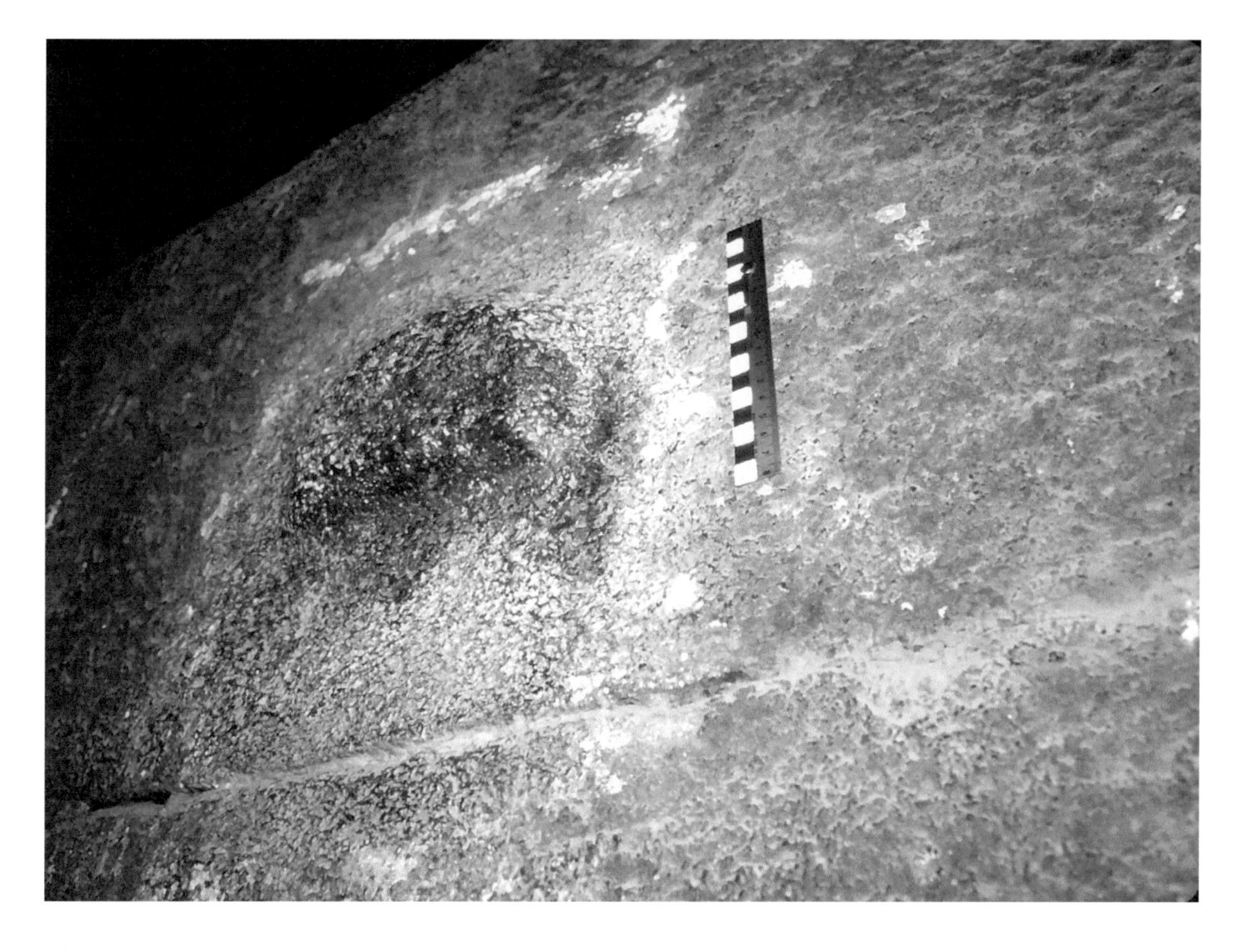

Fig. 128 The boss on the upper slab in the portcullis room. My scale is 16 cm long.

Much has been made of this boss by certain writers. Some say it embodies proportions from which the overall dimensions of the Great Pyramid can be obtained. One writer said that it is not only the key to various measures (Primitive Inch, Royal Cubit, Sacred Cubit) of the whole pyramid, but that it is also a sign that "speaks of the Messianic presence," i.e., Jesus. Of the former assertions Petrie said of the boss and the rough hammer-dressing around it "Anything more absurdly unsuited for a standard of measure it would be difficult to conceive." The "Messianic" proposal is simply ridiculous.

The portcullis room then looked as in figure 129.

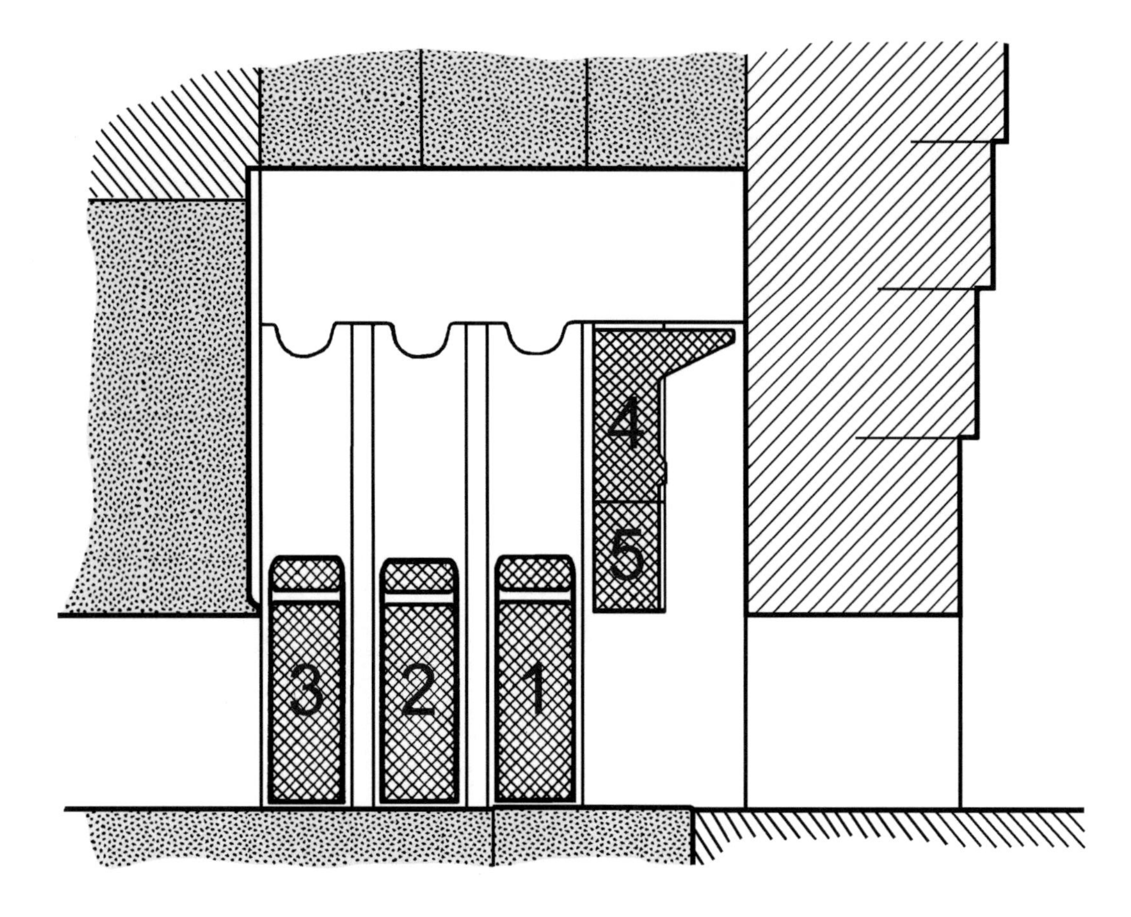

Fig. 129 The portcullis room in its final (sealed) configuration.

The operation of the portcullis room I have just proposed, though seeming complicated by my lengthy description, was not complicated. The entire procedure could have been accomplished in a few hours.

In the scenario I have envisioned, tomb robbers would have destroyed the overhanging part of slab 4 and cut the channel in the southern wall of the portcullis room that I have labeled "broken edge" in figure 123, allowing them to bypass the three main gates.

The King's Chamber

From the south end of the portcullis room a short shaft leads to the King's Chamber (**KC**), a magnificent room faced with large blocks of red granite.

Fig. 130 The King's Chamber and sarcophagus.

The joints between these blocks are so narrow that if their edges were not slightly rounded and chipped they would be hard to detect. How the builders could have accomplished this fine jointing is hard to imagine. It is not apparent that mortar was used (or needed) between the blocks. The walls have been polished by stone rubbing, which must have been extremely tedious work.

Fig. 131 The joints around the duct on the north wall of the King's Chamber.

Khufu's lidless sarcophagus sits in isolation near the west wall of the vault. It is much chipped by souvenir hunters. On the north end of the granite coffin one can see shallow grooves that suggest a saw-cut surface (recall figure 33). Petrie proposed that this sarcophagus was shaped with a jewel-toothed saw, nine feet long.

The "star" ducts

The north and south walls of the crypt each contain a hole that communicates with the pyramid exterior. These shafts are about twenty-two centimeters (9 inches) wide by fifteen centimeters (6 inches) high. They penetrate the granite walls horizontally for about 1.5 meters (5 feet) then angle upward. Both meet the face of the pyramid near the level of the 104th course (81 meters). We can never know if the shafts perforated the now-missing casing. The angled portion of the southern shaft is nearly straight. The northern shaft makes a westward excursion to pass around the Grand Gallery before returning to its original course.

The ducts were constructed by carving channels in blocks that overlay the flat surfaces on which they rest (Gantenbrink), as shown in figure 132.

In 1872 similar shafts were discovered in the Queen's Chamber by British engineer Wayman Dixon. In like manner to those of the King's Chamber, the southern shaft is straight and the northern jogs westward around the Grand Gallery. But these shafts have an interesting difference. They stop about 13 centimeters (5 inches) short of the interior of the room. Also, they do not extend all the way to the pyramid exterior. In March 1993, German engineer Rudolf Gantenbrink guided his tracked robot "Upuaut 2" up the southern shaft, finding it blocked at a distance of about 63 meters, roughly 13 meters short of the pyramid face. A video camera showed the blockage by a slab of limestone that had two greenish metal (copper?) fittings at its upper edge, causing some to hypothesize that these were handles for a door that led, perhaps, to a hidden chamber.

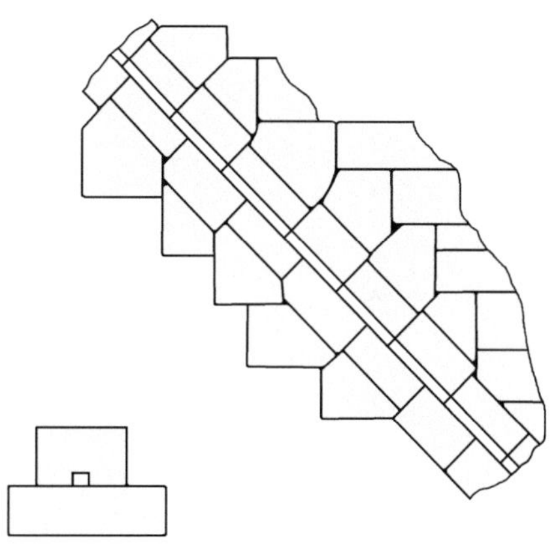

Fig. 132 Construction details of the ducts in G1.

In 2002 a team sponsored jointly by the Egyptian Supreme Council of Antiquities and the National Geographic Society used a new robot to drill a small hole through the slab. Beyond the slab was a 21 cm space, and beyond that space was another limestone plug. This plug was covered by some kind of screening or fibrous mat. The same team was able to send their robot all the way up the northern Queen's Chamber shaft, finding a blocking slab similar the one in the southern shaft, copper handles and all, at about the same distance. As far as I know, no further exploration of these shafts has occurred.

Gantenbrink made laser-assisted measurements of the angles of all four shafts with the help of his little robot. Only one angle, that of the southern King's Chamber shaft, was in close agreement with measurements made by Petrie over a century earlier. I will give only Gantenbrink's values because

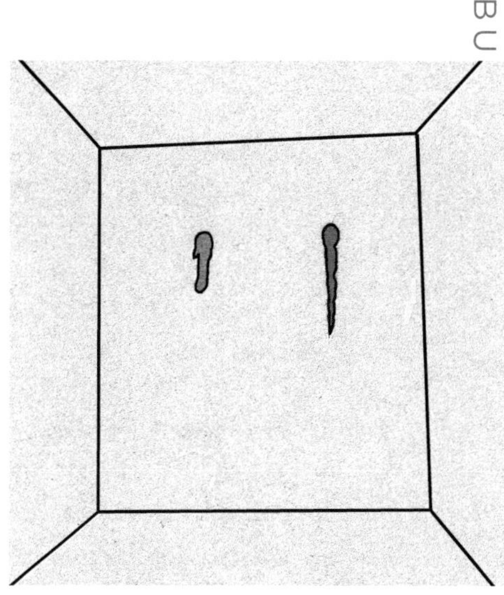

Fig. 133 Blockage at end of southern duct of Queen's Chamber.

he had the advantages of using modern equipment and taking measurements over greater distances than did Petrie:

Shaft	Angle
South King's Chamber	45.0°
North King's Chamber	32.6°
South Queen's Chamber	39.6078° (avg)
North Queen's Chamber	33.3°-40.1° (first 8 meters)

Gantenbrink said of the southern King's Chamber shaft that the 45 degree value was for the long, undamaged part of the shaft which begins a few meters beyond the initial horizontal portion. He also found that this shaft does not run exactly parallel to the north-south axis of the pyramid, but makes several slight bends (east and west) along the way. He said "This effectively disproves Robert Bauval's Orion theory." I will get to Bauval's theory in a moment.

As to the purpose of these ducts there has been much speculation.

Some writers suggested that the shafts provided ventilation for the workers finishing the interior chambers while the upper part of the pyramid was under construction. But the shafts emanating from the Queen's Chamber were never connected to it. That suggests movement of air was not their function.

Others believe these ducts were made to allow access to the king's mummified body by his *ka,* sort of a spiritual double. The Egyptians thought that the *ka* left a person at death, free to travel around the universe and return to the body at will.

In 1964 Virginia Trimble and Alexander Badawy suggested that the King's Chamber shafts were aligned with the meridian transit (highest altitude) of certain prominent stars. Specifically, the southern shaft was aligned with one of the three stars in the belt of Orion, and the northern shaft was aligned with Alpha Draconis, the star closest to the celestial north pole c. 2500 BCE.

A brief astronomy detour:

A *meridian* is a great half-circle on the earth's surface that runs from south to north poles. On the earth's surface these great meridian arcs are measured in degrees of *longitude* starting at Greenwich, England (the *prime meridian*) and measured east (+ degrees long.) or west (- degrees long.).

The second of two measurements needed to locate a point on the earth's surface is its *latitude*. Latitude is the angle from the equator to the position of interest with the earth's center as its apex. If the position is north of the equator it is measured in degrees of north latitude and if south of the equator in degrees south latitude. From an observer's point of view, the angle from the horizon to a particular star is called its *altitude*. A star that appears to rise in the east and set in the west travels along an arc that reaches its highest altitude at the point where it crosses the meridian of the observer.

Now comes the good part. Because the rotational axis of the earth swings around like that of a spinning top (called "precession") with a cycle of about 26,000 years, a star's altitude when it crosses an observer's meridian changes depending on the date on which it is being observed. Thus, if we suspect that Khufu's builders intentionally aligned these shafts with specific stars we can easily calculate the approximate date (or dates) in history when the alignment occurred.

But how can we know if star alignments were intended by Khufu's builders? Robert Bauval, a construction engineer, and co-author Adrian Gilbert, in *The Orion Mystery* (1994), furthered the idea of Trimble and Badawy by noting that the three stars in Orion's belt correspond not only to the positional layout to the three main pyramids at Giza, but also in brightness to the pyramid sizes as well. This seemed to Bauval to be too improbable for coincidence, especially knowing the importance of the Orion stars, and others, attributed by religious texts found on crypt ceilings of Fifth Dynasty pyramids.

Bauval and Gilbert proposed that the left-most star in Orion's belt, Zeta Orionis, popularly called *Al Nitak*, corresponded to Khufu's pyramid. They proposed further, that because this star had an altitude at meridian transit of exactly 45 degrees in 2450 BCE, that that date is one of two likely dates for the construction of Khufu's pyramid, the other being 8,000 years earlier.

Bauval and Gilbert expanded their Giza alignment idea to the "Star Correlation Theory." They say that not only are there celestial equivalents for the three Giza pyramids, but for other Fourth and Fifth Dynasty pyramids as well. They also believe the star match-ups include a correspondence of our Milky Way galaxy to the Nile River. In other words, the stars that correspond in position to these pyramids are also oriented to the Milky Way as the pyramids are to the Nile.

I find this master plan hypothesis not as compelling as the Giza pyramid alignment.

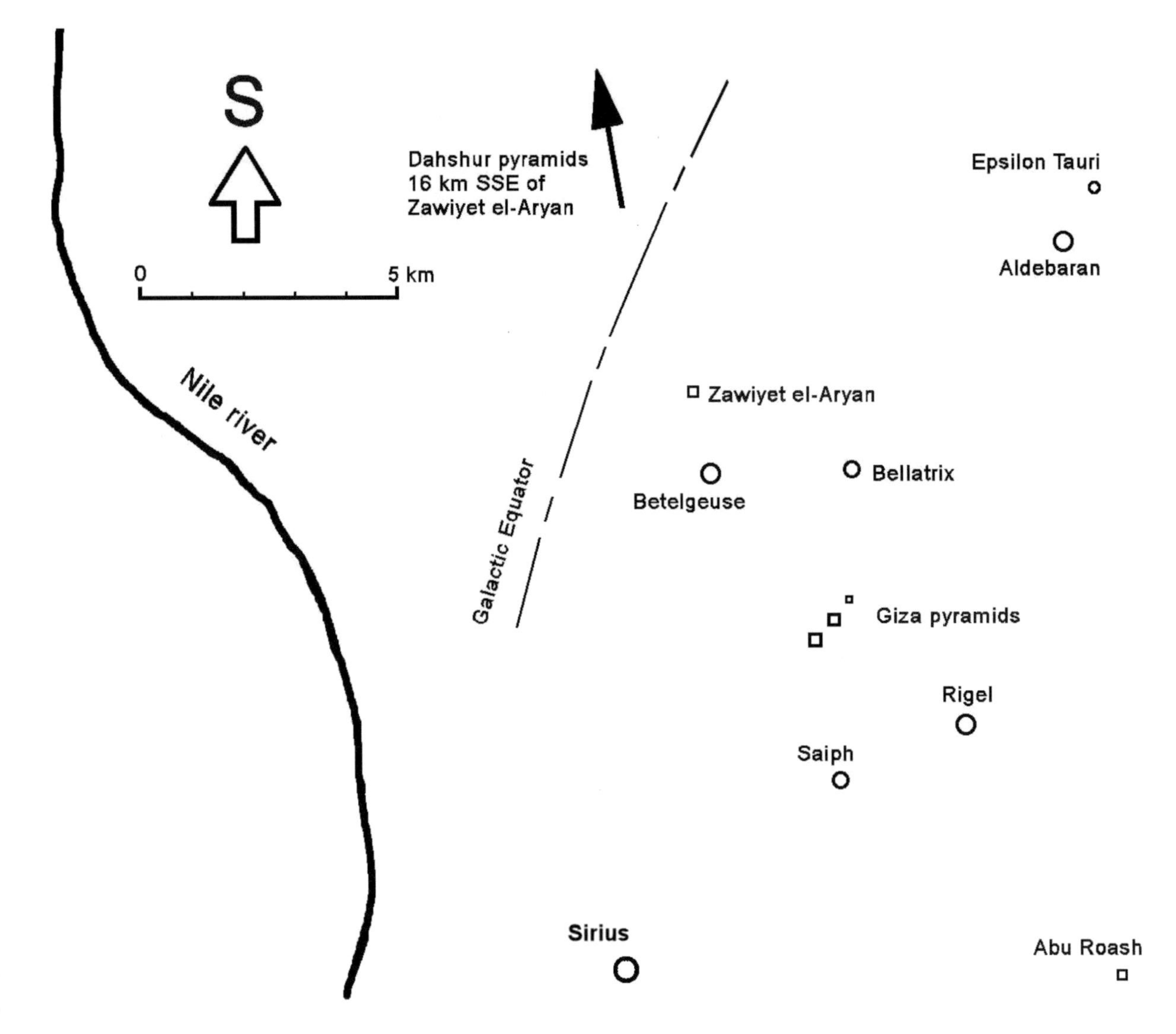

Fig. 134 Star alignments with the Fourth Dynasty pyramids.

In figure 134 I have attempted to show the pyramids (as squares) in the vicinity of the Giza group as their pattern relates to the stars (as circles) in the vicinity of the Orion constellation using the three stars in the belt of Orion and the three Giza pyramids as a common factor of scale. Bauval and Gilbert equate the pyramid of Djedefra at Abu Roash with the star Saiph (Kappa Orionis), but as you can see in the diagram, these two features are not close to matching. Same for The Unfinished Pyramid at Zawiyet el-Aryan – it does not coincide with the position of the star Bellatrix (Gamma Orionis). Further, Sneferu's two pyramids at Dahshur, the Bent Pyramid and Red Pyramid, while they could be seen as having the same positional relationship to each other as the relationship of the stars Aldebaran and Epsilon Tauri have to each other, have no relationship to the Giza group other than being generally south of Giza. Also, Aldebaran, a first magnitude star, is much brighter than Epsilon Tauri, a star of third magnitude, whereas the Red Pyramid is only slightly larger than the Bent Pyramid. Thus the size correlation that appears to hold for the Giza pyramids does not hold for those at Dahshur. As for the Nile corresponding to the central plane of our galaxy, represented in my diagram by the galactic equator, the Nile is some thirty degrees off in angle and substantially farther away from where it should be in the star correlation theory.

For the largest stars in or near the constellation of Orion that might have equivalents near Giza, we see no pyramids. Betelgeuse and Rigel are considerably brighter than the stars in Orion's belt, and nearby Sirius, the most luminous star in the sky, is brighter still. Thus we should expect to see even larger pyramids in these positions, but we do not. Bauval says of this discrepancy "I could only conclude that these [pyramids] had never been built or that they had long since been demolished and had disappeared under the sands of the Western Desert." Noting that these three pyramids should have been substantially larger than Khufu's, this is not a satisfying proposal.

The apparent correspondence of the three Giza pyramids with their stellar counterparts in Orion's belt, I agree with Bauval, would be a remarkable coincidence, even more so when we consider that the southern air shaft of Khufu's pyramid pointed at what would have been the pyramid's celestial correlate at the time the pyramid was erected. If true, the implication of the "Giza Plan" is striking. It would mean that not only were the *positions* of the three main Giza pyramids set from the beginning, but their *sizes* too. Further, this degree of planning would infer continuity counter to the traditional view that each king in the pyramid age was concerned mainly with building his own monument(s). Perhaps the science department (the priesthood?) provided that continuity. I don't care to speculate further.

The ceiling of the King's Chamber

Without doubt the most unusual feature of the King's Chamber is its roof. Gazing upward a visitor sees only a flat surface composed of nine slabs that traverse the narrow dimension of the room. The slabs are enormous granite monoliths that weigh up to seventy-five tons

(68,000 kg), almost twice the weight of a fully loaded 18-wheel semi-trailer truck, which weighs 40 tons.

There's more.

During his 1764 exploration of the pyramid, Nathaniel Davison (1736-1809) noticed a tunnel opening just under the ceiling of the upper end of the Grand Gallery. The intrepid Englishman lashed several ladders together and courageously ascended. The tunnel, deep in bat dung, led to a space directly above the ceiling of the King's Chamber. Here he was surprised to find that the roof of this low room was comprised of another row of huge granite beams.

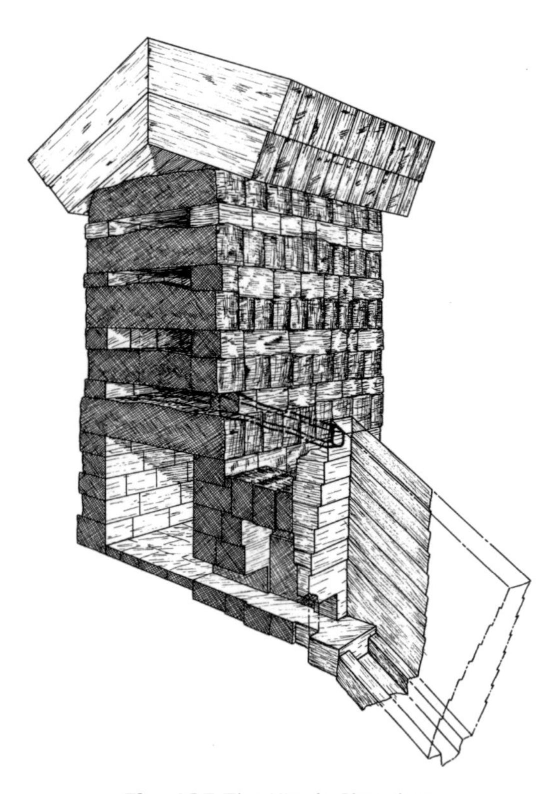

Fig. 135 The King's Chamber.

Seventy-three years later, Vyse hired local quarrymen to tunnel upward from "Davison's Chamber" using blasting powder. He found four more chambers, three spanned by granite rafters, and the last covered with a limestone arch comprised of eleven sets of gabled beams like those protecting the Queen's Chamber. There are forty-three granite rafters in all.

The architect apparently planned the superimposed rooms as proof against earthquakes or gradual settlement of the masonry. If one or two of the uppermost ceilings collapsed, the others might hold. Though nothing has fallen, the King's Chamber has obviously been subjected to tremendous stress. I mentioned earlier that all the beams of the crypt ceiling are cracked through. Petrie found that the beams of *all five* granite ceilings are either cracked through on their southern undersides and northern upper sides, or pulled slightly out of the walls that contain them. He concluded that the southern end of the chamber had dropped "a matter of an inch or two" (Petrie 1885).

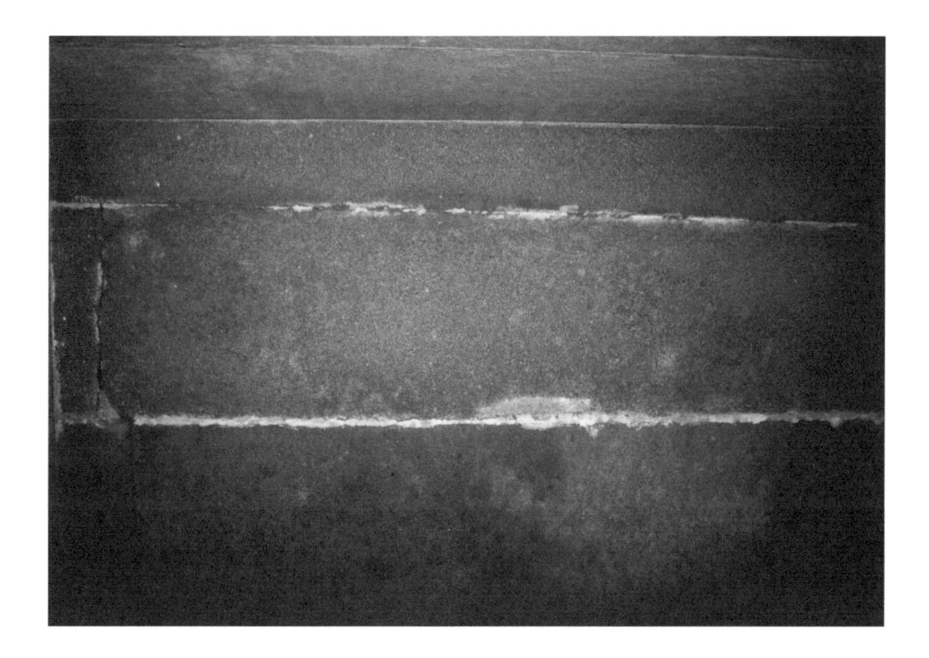

Fig.136 The cracked roof beams at the east end of the King's Chamber. South is to the left.

Why are these beams cracked? The answer can be found by referring to figure 137:

If the southern wall drops a small amount relative to the northern wall it's as if you were to push up with a force as shown on the underside of the north end of the beam. That force creates a bending stress where the beam meets the south wall, a stress that translates to a tensile (stretching) force at point A. The same applies at point B on the upper side at the opposite end of the same beam.

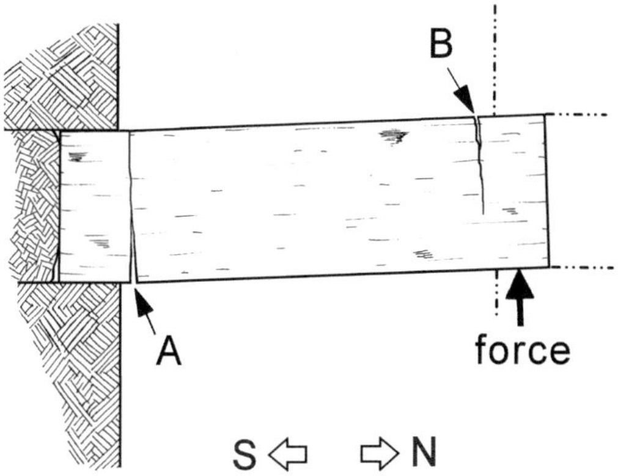

Fig. 137 Cause of cracks in the granite ceiling beams.

Petrie noted that the cracks in the roof beams (figure 136) have been plastered over, an operation probably performed by the builders. Perhaps they wanted to see if the cracks would continue to open. It was also probably they who dug the tunnel to Davison's chamber to check the condition of the next higher ceiling. That the beams had been cracked before the pyramid was sealed shows the difficulty in constructing a room under a nearly three-hundred-foot column of stone blocks.

Because evidence relating to construction time (to follow shortly) has been found in the chambers above Khufu's crypt I will conclude this segment with a discussion of one of the recurring questions concerning pyramid construction: How long did it take to build a pyramid such as Khufu's?

Pyramid construction time

In 1975 the traditional view Dr. Goedicke gave me was that when a new pharaoh came to power in the pyramid age he shifted all tomb-building resources from his predecessor's monument to his own. Thus, there was no overlapping of pyramid construction between succeeding reigns. This tidily explains why we see unfinished pyramids sprinkled among those that are complete or nearly so. Hence, we could arrive at a rough approximation of maximum construction time if we divide the sum of the lengths of reign of the pyramid owners by the number of pyramids they built. Let's look at just the geometrically regular pyramids of Dynasty 4. I will use the years of reign given by Hawass (2006, 9). See table 4.

Table 4 - Years of Reign and Pyramids Built, Dynasty 4.

King	Yrs of reign	Pyramids	Completion
Sneferu	24+	Meidum	Completed
		Dahshur: Bent	Completed
		Dahshur: Red	Completed
Khufu	32*	Giza: G1	Completed
Djedefre	8	Abu Roash	Unfinished
Khafre	26	Giza: G2	Completed
Menkaure	18	Giza: G3	Completed

* estimate by Hawass (2006, 104) based on "a newly discovered inscription found in the Dakhla Oasis in the western desert." It records an expedition sent by Khufu in the twenty-seventh year of his reign.

The six largest pyramids (Sneferu thru Menkaure, omitting Djedefre) were built over a span of 100 years. From this we could say that a pyramid took, on average, about seventeen years to complete, *if* we do not consider size differences or overlapping construction.

Building times change if there was overlapping construction, as Mendelssohn proposed. I think this must have occurred, especially at the same location. For example, at Dahshur, Sneferu's northern pyramid (the Red Pyramid) was probably begun before his southern pyramid (the Bent Pyramid) was completed. It is unlikely that Sneferu, aware of the faults that led to the modification of his southern pyramid, and thus already planning to build another to the north, would have let his building force disperse as his southern pyramid neared completion. Likewise at Giza. It is almost inconceivable that Khafre's pyramid, final ownership perhaps undetermined at first, was not underway before Khufu's building was finished.

Consider. When Khufu's pyramid had reached half its ultimate height, seven-eighths of its volume was established. Its top surface presented only one-fourth the working area of when the pyramid was started. Manpower needs were diminishing at an ever increasing rate. To let the talents and organization of the building machine atrophy, knowing they would soon again be needed doesn't make sense. It's likely that excess workers were assigned to the next great project, G2.

For reasons we don't understand, Khufu's immediate successor, Djedefre, did not choose the large pyramid G2 as his eternal monument. Instead he commenced a much smaller one at Abu Roash, eight kilometers (five miles) north of Giza. This puzzling interlude remains a mystery. More on Djedefre in a little while.

Because of the wide variation in pyramid size the concept of *average* building time is not helpful. Is pyramid *volume* a good basis for comparing building times? No, *work* is a better means of comparison because, in addition to volume, it accounts for the heights to which stones must be raised.

The minimum amount of work, or energy, required to raise an object from the surface of the earth is the product of its weight and the distance it is raised. I will say more about work and energy in chapter ten, but for now it is enough to know that we can compare the *minimum* work required to raise each pyramid, regardless of the stone-raising method. The minimum work on each pyramid can be expressed as a *relative* difference of work required. You can see in table 5, where volume and work on each pyramid are compared to Khufu's pyramid, that relative work makes a greater difference than relative volume.

Table 5 - Relative volume & work in Dynasty 4 pyramids

Pyramid	Relative volume	Relative Work
Meidum	0.263	0.165
Bent	0.599	0.476
Red	0.730	0.548
Khufu	1.000	1.000
Djedefre	0.045	0.007
Khafre	0.912	0.893
Menkaure	0.098	0.044

Notice that Sneferu's three pyramids, while together possessing 59 percent more volume, required only 19 percent more work than did the pyramid of Khufu. Notice also that only 4.4% of the work required to raise Khufu's pyramid was needed for Menkaure's pyramid, and less than 1% was expended on the unfinished pyramid of Djedefre.

Of course these figures ignore not only the efficiency of the stone-raising method, but also the auxiliary labor of quarrying blocks, transporting them to the pyramid, moving them across the pyramid, ramp building, and excavation of underground chambers.

Now let's take the relative work needed for these pyramids and apportion them to the span over which they were built. I'll use 100 years as before.

Table 6 - Pyramid work in proportion to 100-year span.

Pyramid	Relative work	Apportioned build time, years
Meidum	0.165	5.3
Bent	0.476	14.5
Red	0.548	17.7
Khufu	1.000	32.3

Pyramid	Relative work	Apportioned build time, years
Khafre	0.893	28.8
Menkaure	0.044	1.4
	Total	100

I ignored Djedefre's uncompleted pyramid at Abu Roash, as its work is not significant.

I could account for the mismatches in lengths of reign to pyramid building time if the following assumptions were true:

1. Sneferu's Bent Pyramid was started a few years before his Meidum pyramid was finished.
2. Sneferu's Red Pyramid was started a few years before his Bent pyramid was finished.
3. Sneferu reigned more than 24 years (Hawass says up to 54).
4. Djedefre spent much of his reign finishing Khufu's pyramid.
5. Menkaure spent a portion of his 18-year-reign finishing Khafre's pyramid.

These premises would explain Djedefre's apparent lack of production for his eight year reign. Same for Menkaure. They would also yield a more credible allotment of time for building Khufu's pyramid than the 20-year period Herodotus received as hearsay 2,200 years after the fact.

Another indication of the required building time: In the spaces between the multiple ceilings above Khufu's crypt, Vyse found layout lines and other markings of the builders. Among these markings are "year 17" along with Khufu's name. Hawass says (2006, 40) "The Egyptian government took a census of cattle every two years, and kings of the early periods numbered their reigns according to these censuses. Though we know there were several annual censuses during Sneferu's reign we cannot simply multiply the count by two."

Let's say these stones were installed in the seventeenth year of Khufu's reign. They lie about fifty-two meters (170 feet) above zero level, so the pyramid's *volume* was seventy-three percent complete. But as I explained above, the minimum *work* needed to raise a stone is the product of its weight and the height it is raised, so at this elevation only 44 percent of the work was complete.

Because of the need to measure and prepare its foundations, and construct the causeway to the Nile from which the casing stones were brought, the pyramid probably didn't start to rise before Khufu's second year. The crypt ceiling blocks would have been placed, then, about fifteen years after the building started. A straightforward proportion (15/.44) then yields thirty-four years as the total time to erect Khufu's pyramid.

Robert Bauval and Graham Hancock, in *The Message of the Sphinx* (1996, 102), suggested that the "year 17" and Khufu's name graffiti were forged by Col. Howard Vyse, though they did not offer a guess as to his motive. That allegation has been rebutted by Hawass, who made observations using strong lighting in these chambers. Hawass found that many of these inscriptions extend into spaces which cannot be reached today, and thus could not have been reached in Vyse's era (Lehner/Hawass interview, NOVA, 1997). Hence it is near-certain that the block inscriptions are ancient.

Another possibility regarding the references to Khufu in these chambers was offered by Robert Schoch in his book *Pyramid Quest* (2005). Schoch examined the markings and concluded they were made by the builders of the pyramid. How to preserve his theory that the Great Pyramid predates Khufu's reign by several thousand years? He suggested the Great Pyramid already had the name of Khufu when Khufu came to power, and he took the name of the pyramid for himself!

The masonry courses of G1 contain interesting cycles of thickness variation, perhaps another indication of building time. Going upward we periodically encounter a thick course followed by successively diminishing ones. This characteristic was first reported by Petrie on behalf of Mr. J. Tarrell.

Today G1 has 201 full courses plus two partial courses that form a central mound on the top surface. Greg and I measured course thicknesses as we climbed the southwest angle in 1978. I plot our measurements in figure 138.

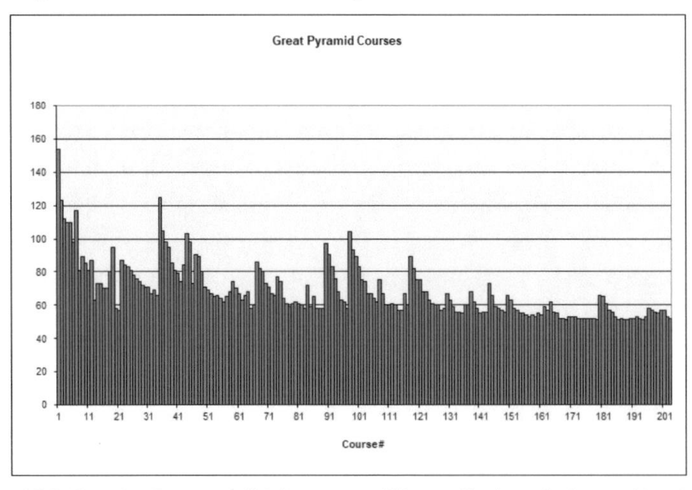

Fig. 138 Graph of course thickness on G1, vertical scale in centimeters.

Because the backing stones are not perfectly level, and because we rounded each measurement to the nearest centimeter, the measurement of each course could be in error by plus or minus one centimeter, perhaps a little more. That would result in a total error of at least plus or minus two meters in the sum we obtained for the height to the top of course 201, 137.53 meters (451.21 feet). It is also possible, even likely, that settlement of the building over the span of forty-five hundred years has slightly reduced its original height. A survey using a laser rangefinder could tell us for sure what the present height is.

Petrie and Tarrell identified nineteen cycles of course-thickness variation. They attributed the cycles to periods of major resupply of casing blocks from the quarries. They also found that the area of the square at many of the levels where a thick course begins (the base of the thick course) is proportional to the base of the pyramid in ratios of x/25, where x = 2, 4, 6, 7, 10, 14, and 16. The values that Greg and I found tend to confirm these relationships. These ratios, then, would seem to be planned rather than coincidental. For what purpose? Bookkeeping? I don't know.

Petrie presented a hypothesis that included correlation of these cycles with the twenty year building time given by Herodotus. His theory is weakened, though, by his incorrect assumption that course thickness is constant from the outside through the interior of the pyramid.

The even courses that we see on the outside of the pyramid do not retain constant thickness as they extend into the building. You can see in the large hole in the south face of Khufu's pyramid that

Fig. 139 West side of large notch on south side of Khufu's pyramid.

the heights of backing and packing stones are matched for only a few meters into the body of the pyramid.

The supply cycle from the Tura/Maasara quarries may have affected course height, but not the pace of erecting the building. Tura limestone comprised a relatively minor four to five percent of the pyramid, but was much harder to come by than local stone. The casing material was probably transported only during high Nile when barges could be unloaded at the quay in front of Khafre's Valley Temple.

The masons didn't need to accumulate ten years' worth of stone before the building could begin, as Petrie proposed. But they did have to have enough casing of a given thickness to complete a particular course. This they would have ascertained before the course was started.

It is likely that casing blocks were sorted by thickness in a building yard close to the pyramid. The masons seem to have preferred thicker blocks if available. The variation in course thickness may have resulted because the builders, who could count and measure, periodically lacked enough casing stones of a thickness I shall call "fat" to complete the next course. One or two courses higher the block count needed to complete the reduced perimeter had shrunk so that the fat stones could be used.

Taunting death

Before leaving Khufu's pyramid I will confess a foolish act. On our second trip to the top of the pyramid Greg and I spent about an hour taking photos and mapping the top surface. We decided to descend via the northeast angle. Greg started down. I wanted to see if I could find possible machine evidence on the north face of the pyramid. As the steps didn't seem too high (only 22 inches), I began stepping down the midface while facing *away* from the pyramid. After about four steps, a foot of my camera tripod, the top of which was tied at the top of my backpack, caught the edge of the block behind me. The tripod pushed me away from the pyramid.

Three things occurred to me in the next instant.

First, the face angle of the pyramid, at fifty-two degrees, is steep, so if you fall at the top you will not stop tumbling until you reach the bottom. And because the human head is not built to resist repeated impacts on rock, one or two tourists die every year in this dreadful way.

I also knew that I was being forced far enough from the pyramid that I could not land on the next step down. I would have to jump to the second step below.

The third thing I knew was that my landing would have to be with enough resiliency that I would stick and not bounce. It's amazing how adroit your mind can be when death is imminent.

As I departed the place where I stood I gave a slight twist of my body so I could get my hands and feet positioned for the places where they had to land. My attention was focused on those two spots.

Since I am here to tell the story, I must have stuck.

I sat and contemplated the vagaries of mortality.

My Nikon SLR camera strapped to my shoulder had struck the stone beside me, making a loud clack. Greg turned and saw me sitting on a block, staring into space. Puzzled, he asked "What are you doing?"

"Give me a few seconds and I'll tell you."

DJEDEFRE'S PYRAMID

Khufu was succeeded by Djedefre, his son or younger brother. Djedefre forsook the Giza plateau to erect his pyramid on a secluded ridge at Abu Roash. This structure, though unfinished, is noteworthy in several respects. It would have been much smaller (13 percent by volume) than Khufu's pyramid, it was being raised in the form of large steps, and there are no traces of construction ramps.

Djedefre will probably be remembered most for his work at Giza. It was he who entombed the two wooden ships of Khufu, one of which is now displayed in the museum on the south side of the Great Pyramid.

KHAFRE'S PYRAMID

Khafre, son of Khufu, followed Djedefre as King of Egypt. He ruled for thirty years. Khafre built his pyramid (G2) just southwest of his father's monument. Having a base length of 410 cubits, G2 is slightly smaller than G1. Its sides slope with a *seked* of five-and-a-quarter, making it a little steeper.

Fig. 140 Khafre's pyramid, looking northeast.

Though G2 is only marginally smaller than G1 it has attracted much less attention. The reason is probably the simpler layout of internal apartments. Still, inside and out, it has many intriguing details.

Khafre's pyramid has two entrances. One is in the usual position on the north face, the other is just below, descending from the court in front of the pyramid.

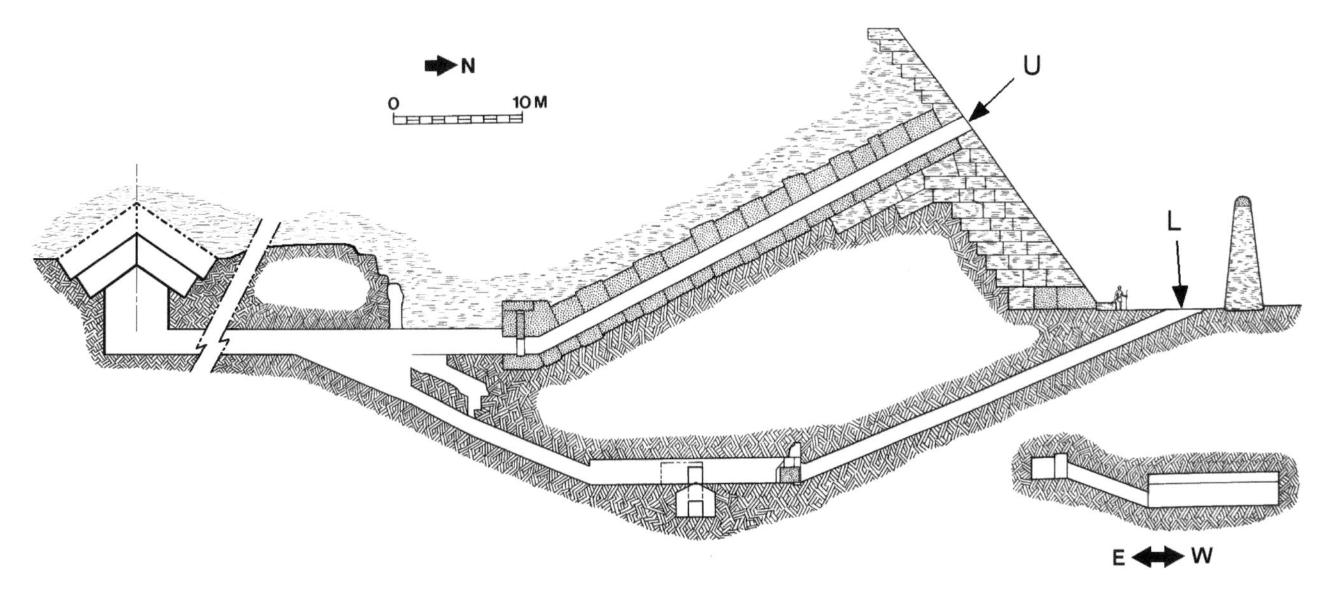

Fig. 141 The internal layout of Khafre's pyramid.

The precedent for using two entrance shafts had been set in the Bent Pyramid. But unlike the Bent Pyramid's corridors, which lead to separate chambers, G2's entrances lead to the same room.

Eleven meters (36 feet) above the pyramid base, the upper shaft (**U**) descends at a slope of 27 degrees until it meets a long horizontal corridor leading to a large vault roofed with butting limestone beams. At the west end of this room, surrounded by limestone blocks, is an empty sarcophagus of red Aswan granite. The sarcophagus and the central location of this room suggest that this chamber was to be Khafre's burial chamber.

Fig. 142 Red granite sarcophagus in Khafre's pyramid.

Lower shaft (**L**), completely carved in bedrock, descends to a horizontal portion from which an intersecting passageway further descends to a long room extending to the west. What was the purpose of this long, skinny room? Was it a storeroom for funerary goods? Was it the correlate of the Grand Gallery in G1? Who knows?

The roof of Khafre's crypt is similar to that of the Queen's Chamber in G1. But Khafre's vault is more strongly built. The entire room, save the ceiling, is hollowed out of bedrock. The foundations of the gabled beams are not built-up masonry. They are the same limestone bedrock from which the room is carved. The upper layer of butting beams in figure 141 is my guess.

The interior features of Khafre's pyramid are so solidly constructed that it seems lessons were learned from Khufu's building. For the crypt roof Khafre's architect did not choose the *latest* design. That would have been Khufu's simple-beam ceiling. Instead he chose the *best* design, that of the Queen's Chamber. Recall that masons plastered over some of the cracks in the granite beams of the crypt ceiling, so they knew the simple-beam ceiling was flawed. They learned from that, and used the gabled arch for all crypt ceilings thereafter.

An amusing aside: Early one morning Greg and I arrived at Khafre's pyramid to photograph interior details. I led the way. Upon entering the crypt I unknowingly flushed a large bat that promptly decided to leave the pyramid. I heard a commotion accompanied by a loud expletive in the corridor behind me. It was Greg. I asked him what was the matter. "A humongous bat was coming straight for my head. I had to duck to get out of that bastard's way."

Turning now to the outside of the pyramid, for quite a while I was puzzled by an aspect of G2. The courses of backing stones on the upper part of its superstructure, just below the remaining casing, all appear to be the same height and have their outer edges aligned, a combination we don't see in any other pyramid.

The first time Greg and I tried to climb up there to take a closer look we attracted such a crowd of screaming pyramid guards that we had to abort our mission. We needed full concentration for this task. It is strictly forbidden to climb G2 (and since our 1978 trip, G1 also) because of the loose footing and overhanging casing at the top. Not easily discouraged, we arrived the next morning before the guards appeared. We scaled the southeast angle to a point just below the casing. We were comforted to find the overhanging masonry buttressed in places with concrete pillars.

My suspicions were confirmed. The blocks laid bare in this section are not backing stones at all. They are the facing of a carefully assembled nucleus structure having steps two cubits high.

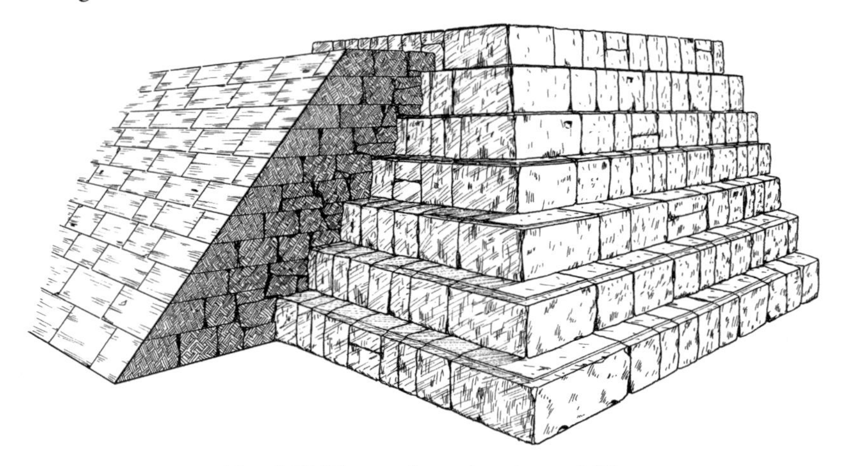

Fig. 143 The nucleus structure of G2.

Fig. 144 Casing and core structure at top of G2.

The casing stones in the preserved upper section vary between 1.0 and 1.8 cubits (0.5 - 0.9 m) thick, and so do not correspond to the nucleus steps. Further, the largest casing blocks in the lower portion of this section must weigh in the neighborhood of four tons. It follows, then, that the stone-raising method was not much restricted by the height blocks had to be raised.

The nucleus blocks had been placed with their outer edges aligned. Many, but not all, have shallow cuts along their outside upper edges that may have been used in leveling operations (denoted by arrow in figure 145). Most of these stones are headers. Many were laid on edge to attain the required two-cubit height.

Fig. 145 Leveling cuts in G2 nucleus blocks.

From the face of the casing to the nucleus, a distance of about three meters, we see, in order, casing stones, backing stones, and rough core-filling stones.

Nucleus structures of the pyramids

G2's nucleus structure is different from those found in other pyramids. As these features are relevant to our understanding of the building process, let's look at them more closely.

After the invention of the geometrically regular pyramid at Dahshur, few pyramids were built in the older stepped form. However, a strange fact is: All later smooth-sided pyramids that are ruined enough to reveal their interior make-up contain a stepped nucleus (a stepped pyramid) within.

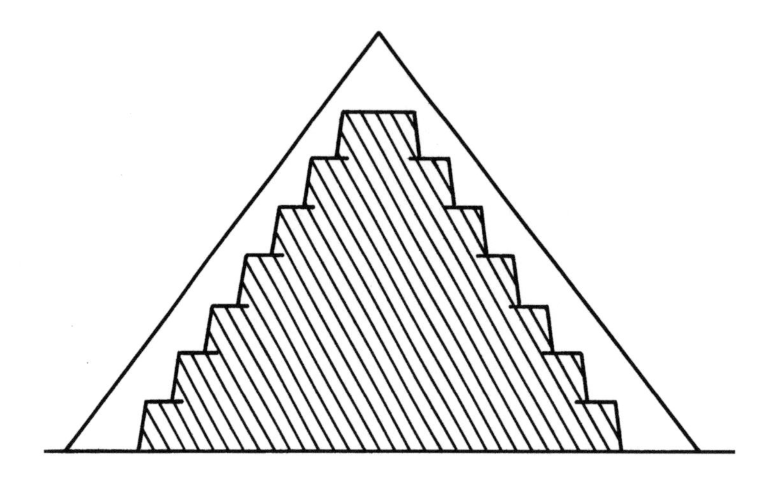

Fig. 146 The internal stepped nucleus.

See table 7. In the pyramids having large internal steps the steps are 10-12 cubits high. The stair-steps of Khafre's nucleus structure are only two cubits high.

Table 7 - Interior Structure of the Pyramids

Pyramid	Nucleus*
Bent Pyramid	X
Bent Pyramid Subsidiary	X
Red Pyramid	X
G1 (Khufu)	X
G1a (Hetepheres?)	X
G1b (Meretyetes?)	large steps
G1c (Hetnutsen)	large steps
Djedefre	large steps

Pyramid	Nucleus*
G2 (Khafre)	stair-steps
G2a (cult pyramid)	X
G3 (Menkaure)	large steps
G3a (Menkaure family)	X
G3b (Menkaure family)	large steps
G3c (Menkaure family)	large steps
Mastabat el-Fara'un	large steps
Dynasty 5 to 8	large steps

* X denotes unknown structure

Why did the builders go to the trouble of constructing a pyramid in steps if it was to be covered with a smooth-faced pyramid? Three possibilities have occurred to me.

1. The builders developed the "formula" of constructing stepped pyramids in Dynasty 3 and couldn't change their method. I reject this because the builders displayed continuing design evolution.
2. The stepped form was a symbolic "stairway to heaven" for the king's spirit. Seems like a lot of extra work for that benefit.
3. The stepped nucleus somehow facilitated or aided construction, even for regular smooth-sided pyramids. This one makes the most sense to me.

If construction was facilitated by building stepped cores we must still determine why this was so. Some possibilities:

a. The stepped nucleus helped maintain accuracy of the final, regular-pyramid, shape.
b. The stepped configuration can be added to; regular pyramids cannot be safely enlarged.
c. The stepped nucleus helped prevent settlement of core masonry.
d. The stepped nucleus could be advanced if casing material was not immediately available, thus saving overall building time.

I think some or all of these reasons could be true.

To elaborate on possibility **d**, pyramid superstructures may have been built by two classes of masons. One class, perhaps seasonally or rotationally employed, built the core and filling blocks. They comprised the bulk of the work force and performed most of the grunt work. The second group, more highly trained and skilled, fitted and laid the casing and backing stones, and were responsible for constructing the internal apartments and corridors.

Returning to Khafre's pyramid, at its northwest angle its nucleus is composed of bedrock to the level of the sixth course. Because the ground slopes down from northwest to southeast the builders had to quarry stone away from the northwest corner and build up the southeast foundations with huge blocks of local stone. According to Selim Hassan (1960, X, 50), one of these blocks measured five by six meters (16 x 20 feet) in plan (top area) by more than five meters (16 feet) deep. Hassan did not say how he ascertained its depth, which makes me a little skeptical, but if true, such a block would weigh nearly 400 tons. I have measured several core stones at the southeast and northeast angles of G2 that attain a hundred tons. These are heavier than any blocks visible in or on G1.

Fig. 147 116-ton backing stone at south end of east face of G2.

One can observe 160-ton blocks on the south side of Khafre's valley temple. These blocks, I believe, were *rolled* into position from the outside. That some possess lever notches at their upper outside edges, and the lack of any other method I can imagine, leads me to this proposal.

Fig. 148 Levering notches (at arrows) on core block, south side of Khafre's valley temple.

Khafre's builders used red Aswan granite for the first casing course on G2. It may have been to strengthen the pyramid's foundation, or maybe it was for aesthetic appeal. Many of these blocks were wedge-split away in the Middle Ages for use elsewhere. A few remain *in situ*. The casing slope I mentioned earlier, a seked of five-and-one-quarter, equates to the ratio of 4/3, and a triangle having proportions of 3-4-5. Some have suggested that this slope shows that Khafre's builders knew that a triangle with these proportions contains a right angle. Though I haven't seen evidence elsewhere, I tend to agree. It seems unlikely that the science department would have failed to notice this relationship.

The upper surface of the granite casing was made level all around the pyramid. The bottoms of the granite blocks were let into the bedrock to varying depths, according to the thickness of each block. The masons found it easier to chisel soft limestone than hammer-pound the granite. The uneven lower edges of the granite blocks were hidden by a border of white limestone pavement. The joints between the casing stones are not noticeably wider than their counterparts in G1.

I measured flatness of the polished faces on a few of the granite casing stones. Each block I checked was flat within a half millimeter over the span of my cubit-long straightedge. This degree of accuracy exceeds that on the blocks of the King's Chamber in G1.

Fig. 149 Measuring flatness of G2 granite casing block.

The granite casing and backing stones are securely joined with gypsum mortar. So well are they bonded that medieval stone thieves were forced to wedge-split them, taking away only the outer portions.

In many places one can see that the granite casing stones could only have been pushed or rolled into position from the outside. Giant core blocks and bedrock have been cut to receive specific pieces. The laying-on of granite casing can also be seen in Khafre's mortuary, sphinx, and valley temples. Note, however, that the sequence of laying casing blocks *last* is the reverse of what the builders preferred. They laid casing blocks *first* when they could easily get to the rear of the casing.

Fig. 150 Cuts in core blocks (at arrows) to receive granite casing of Khafre's valley temple.

THE GREAT SPHINX

Beside Khafre's valley temple to the north is the sphinx temple, and behind to the west is the Great Sphinx itself (refer to figure 7). Carved from an outcrop or raised portion of the Giza plateau, the Sphinx is a monolithic structure having the body of a lion and the head of a pharaoh. Most Egyptologists believe the Sphinx was created by Khafre. In recent years, skeptics have argued that the Sphinx is not contemporary with Khafre. It is much older – dating perhaps to 10,000 years BCE. For one thing, they say the face of the Sphinx does not resemble the life-like face of Khafre on his exquisite diorite statue that was found in his valley temple. They also cite the vertical, water-eroded channels in the body of the Sphinx and the southern and western faces of the quarried trench that surrounds it. They claim that for water to form these deep channels it would take much longer than 4,500 years.

Fig. 151 Water-eroded channels in south wall of the Sphinx trench.

But Lehner and Hawass raised two bits of evidence in rebuttal. First, the quarry face that forms the south wall of the sphinx trench (fig. 152) runs *parallel* to the east-southeast-oriented causeway that connects Khafre's pyramid temple to his valley temple. Because the long axis of the Sphinx is east-west, the proximity of the causeway did not allow as wide a trench at this location as in other places. This suggests that the causeway, a necessary feature of Khafre's pyramid complex, predated, or is at least contemporary with, the carving of the Sphinx.

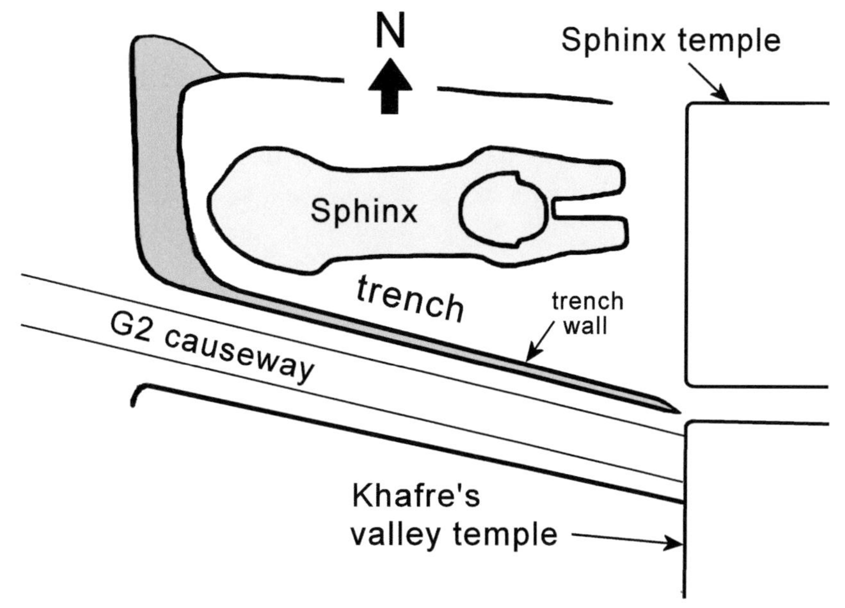

Fig. 152 The south wall of Sphinx trench is parallel to the G2 causeway.

The other indication that the Sphinx belongs to Khafre is that the Sphinx temple is built with the same masonry technique as Khafre's valley temple.

As for the face of Khafre's diorite statue not matching that of the Sphinx, note the broad faces of Rameses II on his megalithic statues at Abu Simbel and compare them with the narrow skull of his mummy. The sculptors of monumental statuary seem to have followed canons, as tomb painters often did, in rendering their subjects without regard to exactness.

Of course, skeptics, not to be thwarted, contend that at least parts, if not all, of the complex attributed to Khafre may be much older than the Fourth Dynasty. Robert Schoch (2005), for example, argues that much of the structures of G1 and G2 were built earlier than the Fourth Dynasty. In the fourth chapter of his book he presents a hypothetical case that "possibly the Fourth Dynasty Egyptians were rebuilding and adding to a much older structure." By the twelfth chapter he has converted his hypothesis to a fact: "What we know as the Great Pyramid was not built all at once, in a single historical episode, but in stages across a long span of prehistory and history."

Khafre's pyramid complex must have been exceptionally beautiful in his time. He preferred strength and simplicity to architectural innovation, achieving an uncluttered integrity that marks, I think, the artistic zenith of the great pyramid age.

MENKAURE'S PYRAMID

The Third Pyramid of Giza (G3) was built by Menkaure, a son of Khafre. His pyramid is much smaller than the pyramid of his father, containing only eleven percent of that masonry. Why it is so much smaller we can only guess. Perhaps Menkaure was following the "Giza plan," mentioned earlier, or maybe he simply commanded fewer resources than did his father.

Fig. 153 The pyramid of Menkaure, looking southwest from G2.

Though small it would be, it would not lack quality of construction. It was to be covered in hard stone of the First Cataract, Aswan rose granite. For some reason only the first sixteen courses were faced with that rugged stone, the rest with white Tura limestone.

Edwards surmised (1972, 119) that Menkaure's death may have prevented completion of the granite mantle. A more likely explanation will be presented in a moment.

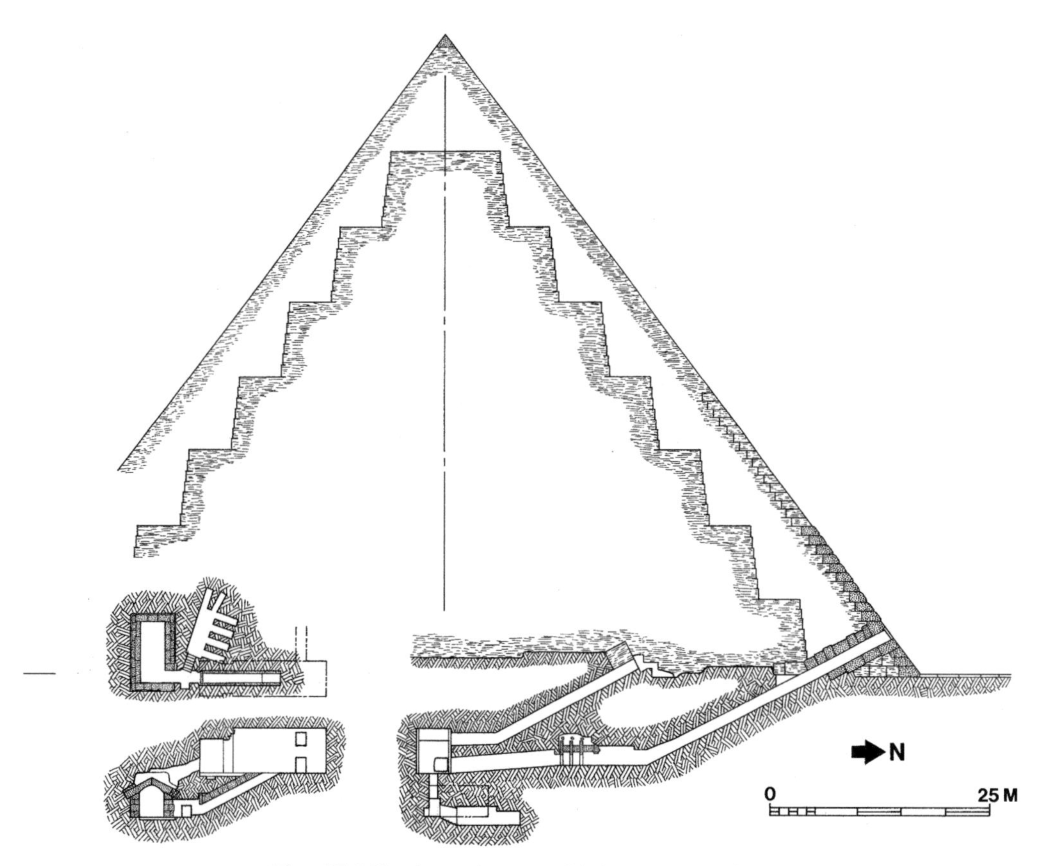

Fig. 154 Menkaure's pyramid, in cross-section.

All but seven courses of granite casing blocks remain on G3. Of the blocks stripped away, many lie in half-buried disarray around the base. That the granite courses numbered sixteen was deduced by Maragioglio & Rinaldi from their observation that corner backing stones above the sixteenth course are large and overlap each other; those below are smaller and do not overlap.

The few Tura stones scattered around the base have been finish dressed. No so for the granite. The granite blocks are unsmoothed except for rectangular patches in the centers of the pyramid's four faces, next to the ground. These areas have been dressed roughly flat, probably by pounding with the hammer-stones I mentioned in chapter two.

Fig. 155 Granite casing on east side of G3. Note the oblique rising joints.

Bosses on the granite casing stones

Some of the granite casing stones have odd-shaped protuberances that some writers have called "lever bosses." Reisner found by experiment, however, that levers cannot gain enough bite on the projections to maneuver the stones. Maragioglio & Rinaldi suggested the bosses are "only the remains of a first working and shaping of the blocks."

Fig. 156 Granite casing, north face of G3. Arrows point to some of the bosses.

With respect to the granite bosses I noted the following:

a. Fewer than one-fourth of the stones have them.
b. The largest blocks, those we would expect to be in greatest need of lever bosses, have none.
c. Though most bosses appear at lower portions of block faces, a few appear at the upper.
d. Some bosses have outer portions broken off.
e. Blocks with bosses are usually grouped together in rows along the same course, but course-to-course the groups seem to have no relationship.

f. Some blocks have two bosses and some have one, the number irrespective of block size.
g. Almost all blocks were laid as stretchers.

My explanation for the bosses: They are the product of dividing large blocks into two or more pieces, thereby stretching a limited supply of precious granite. The masons deemed it safer to separate these blocks by making hammer-pounded channels than by more risky wedge-splitting. My drawing shows two methods of block division that would produce what we see on the pyramid.

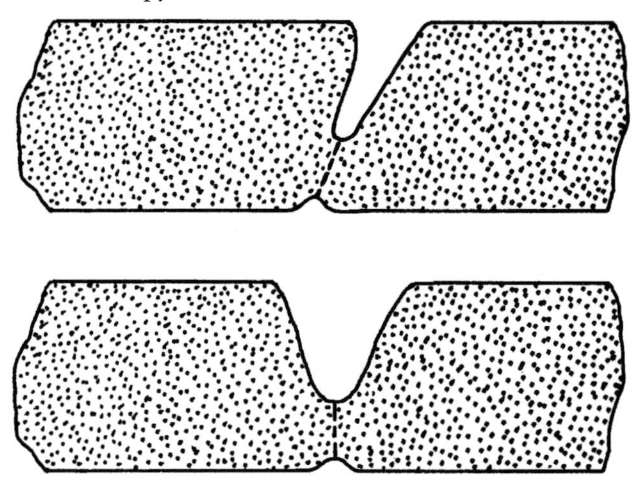

Fig. 157 Bosses (at arrows) produced by block dividing.

This hypothesis not only agrees with observation, but helps explain why Menkaure's pyramid was only partly cased in granite. Either some condition interrupted the supply from Aswan, 500 miles upriver, or the rate of supply was lower than demand could tolerate.

In the middle of the north face of G3 is a large chasm attributed to vandalism by one of Saladin's sons in the twelfth century CE. The rift makes visible a core structure in the form of large steps. This is a good example of a pyramid nucleus that is so well built that it could stand alone and present an excellent appearance.

The step faces of the nucleus are composed of large blocks of local limestone in leveled courses. Each course is set back a small distance from the one below, giving the step faces a slight batter.

The chasm also provides an opportunity to view an extensive area of step-filling in cross-section. Figure 158 is my drawing of the east wall of the rift.

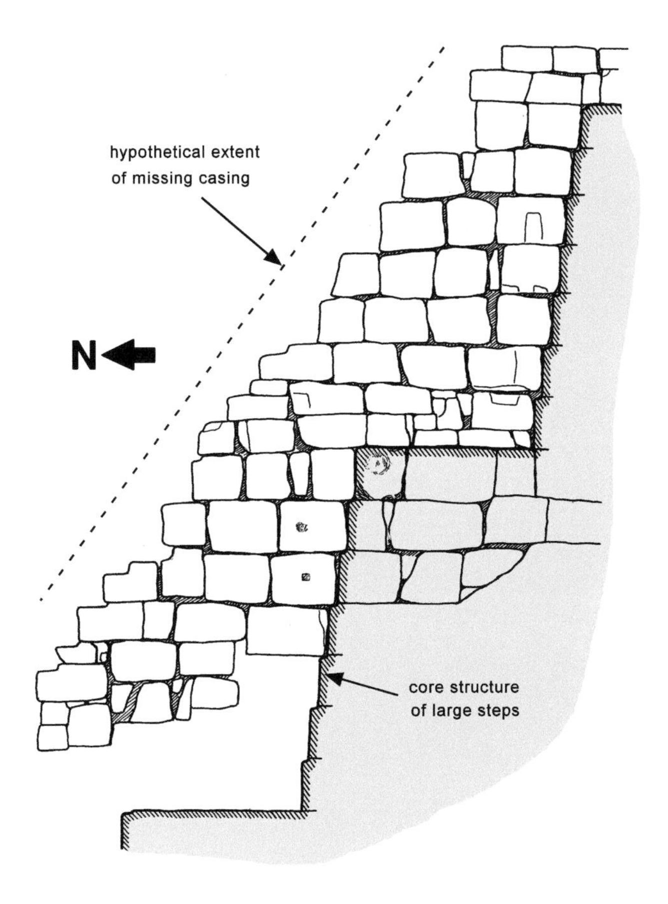

Fig. 158 East wall of G3 chasm.

Points of interest:

a. The blocks laid against the step faces are about the same height and size as the blocks that form the step faces.
b. Block size somewhat diminishes from the step faces outward.
c. The courses of step filling masonry are only roughly distinct.
d. The casing stones were not as thick as the courses of large-block filling.
e. The limestone of the nucleus appears to be the same material as the filling outside of the steps.
f. There are fitting cuts in the outside upper edges of some backing stones.

g. On the faces of nucleus blocks of the core structure, in red ochre paint, are horizontal lines, one cubit apart, along with other marks of the builders.

Fig. 159 Horizontal lines on G3 step face spaced one cubit apart.

Fitting cuts in backing stones were made because each cut becomes part of the bedding seat for a casing stone of the next higher course. Now, as I am touching on a long-running argument, I will make a (probably-futile) attempt to settle it forever.

Casing/filling block-laying sequence

The point of debate is this: In pyramid construction, was casing the last-laid or first-laid portion of each course?

Here's the answer: Where possible, the builders preferred to place casing stones first, then backing stones, and finally the filling behind. To support this contention I offer four observations:

Observation 1 - *The irregular profile of the backing stones on pyramids with missing casing.* Refer to figure 160, the top surface of G1. The irregular perimeter is typical of the rest of the building.

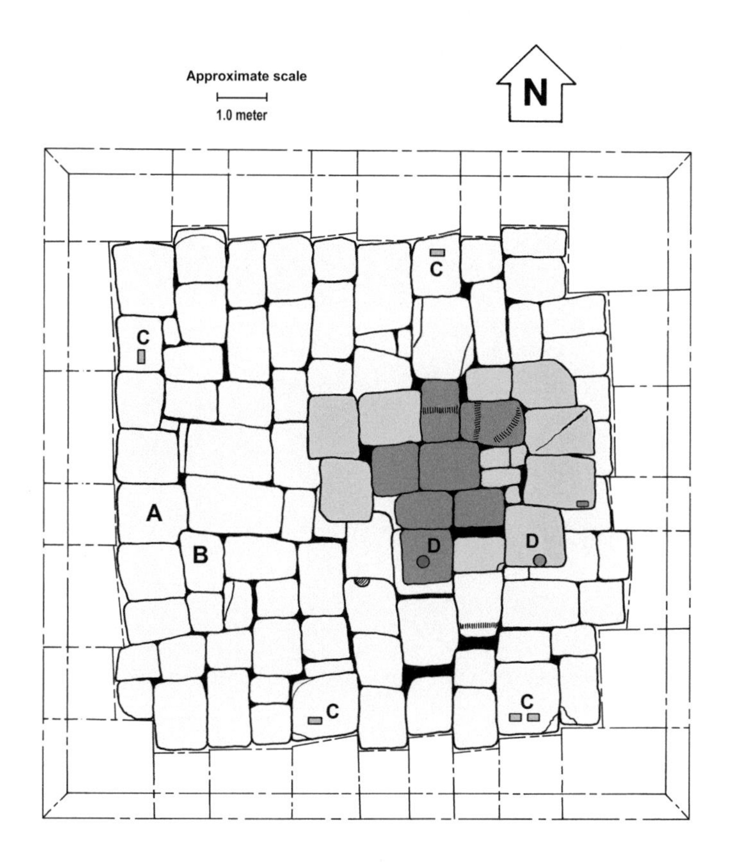

Fig. 160 The present top of Khufu's pyramid, with hypothetical casing in phantom.

I've ventured a reconstruction of the casing, shown by broken lines. The yellow-shaded blocks are in the next higher course (course 202). The darker-shaded blocks are higher still, corresponding to course 203, though most are double height, as can be seen in figure 161. It would have entailed a lot more work to cut and fit the casing stones to the irregular perimeter of the backing stones than to simply fill core blocks behind the irregular casing profile.

Fig. 161 Top of Khufu's pyramid, looking northeast.

Observation 2 - *The absence of vertical fitting cuts in the outside edges of backing stones.* In this context a fitting cut is a vertical notch in the corner of a block. Referring again to figure 160, notice that there are no fitting cuts in the outside edges of the perimeter blocks. There is a fitting cut on the inside edge of core block **A**, which shows that block **B** was placed after block **A**. In other words, block **B** was inserted into the notch of block **A**.

Observation 3 - *Undercut backing stones.* In many places on the pyramid one can see prepared bedding surfaces dressed on the upper faces of backing stones. Sometimes the bedding surface continues underneath the backing stone of the next higher course, leaving a gap, or undercut. Figure 162 depicts the sequence by which undercuts were produced.

Figure 162A shows the leveled seat prepared for the casing stone of the next higher course. Figure 162B shows the casing stone having a roughly-squared outer face now in place. Figure 162C depicts a backing stone placed against the casing, and the casing when finally dressed flat (possibly much later). Figure 162D shows what the backing stones look like after the casing is removed. Note the undercut backing stones. If a casing stone was placed last, that is, if its backing stone was already in place, there would have been no reason (or easy way) to extend the bedding surface for the next higher casing stone under that backing stone. Note: this observation was first made by Olaf Tellefsen in the February 1977 issue of *Construction Dimensions* magazine.

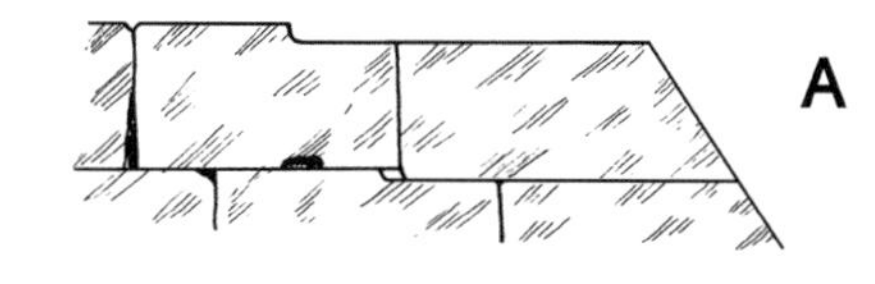

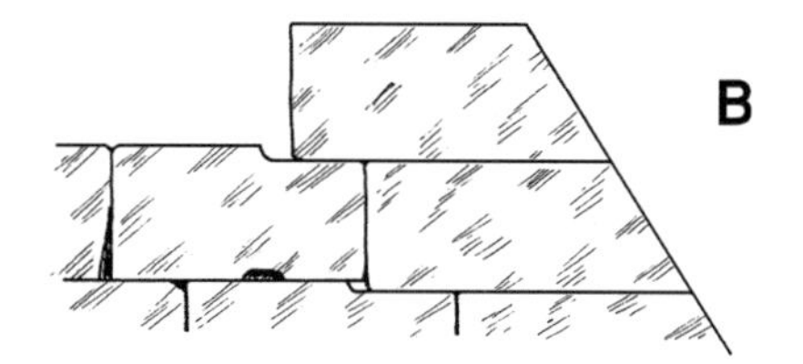

Observation 4 - *The blocks are packed less tight and less precisely from the outside of the Pyramid inward.* In the large rift excavated by Vyse in the center of the south face of G1 one can see (fig. 139) the transition from better fitting of the outer blocks to a rather confused interior. Proceeding inward, courses of even thickness become lost, block size varies considerably, and greater quantities of mortar, chips, and chunks are used to fill voids. The inside is, in fact, a filling put behind the more carefully placed outer blocks.

The order-of-placement I have just described was the sequence preferred by the builders *where possible*. One can notice that there are fitting cuts in first course backing stones on the north base of G1 (figure 186) which indicate that the casing stones were placed *after* the backing stones, the reverse of the rule I proposed above. I think these cuts are exceptional because at these places the bedrock core massif that rises within G1 is close behind the backing stones. Thus, the builders had no choice but to apply the backing stones against the solid rock before laying casing to the outside.

Let's return to G3.

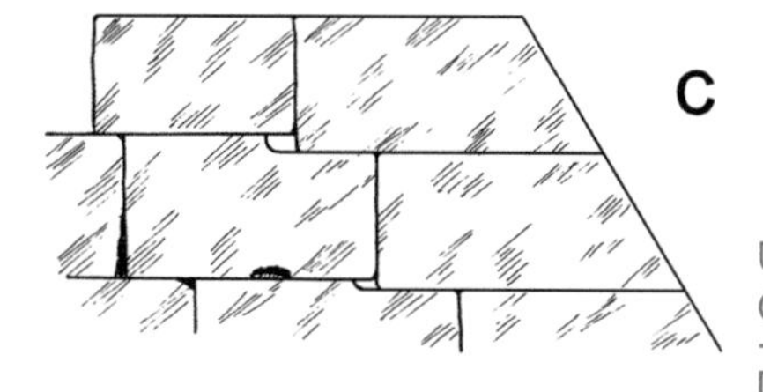

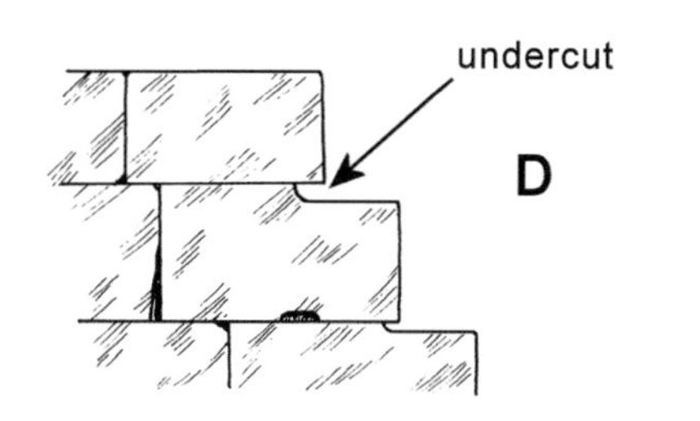

Fig. 162 Casing placement sequence.

With regard to the horizontal lines painted on the step faces, these were presumably used in leveling the courses and marking off major step heights. Leveling lines appear on many walls in other places. What puzzles me is this: Why did the masons mark level lines at every cubit along a steep wall? Why didn't they simply mark every fourth or fifth cubit? Perhaps these lines were being constantly covered by the step filling. If so, then at this pyramid the step filling was only slightly lagging construction of the core.

On the upper portion of G3, fitting cuts tell us the order in which stones were placed. In figure 163, at the southeast angle, the backing stone at top center contains a fitting cut and is undercut. The undercut shows that this block was placed after the corner casing block; the fitting cut shows it was placed before the next casing stone to the west (away from the corner). From these facts it is clear that the masons laid corner blocks first, then added blocks toward the middle of the pyramid face. Today, brick and block masons follow the same sequence.

Fig. 163 Fitting cut leading from SE angle of G3. View looking east.

The interior of Menkaure's pyramid is impressive. Refer back to figure 154. All rooms and corridors except the upper portion of the entrance passage were hewn in bedrock. The entrance shaft descends to a horizontal corridor that is guarded by a portcullis room having three gates, now missing. It continues to a large room that runs east-west.

Also entering the large room, just above the entrance shaft, is another passageway that proceeds northward for a few yards and then slopes upward, abruptly ending where it meets the masonry of the pyramid. It's likely that the granite rafters of the crypt ceiling were brought through this tunnel, their route to the crypt being a down-sloping ramp.

Immediately before the crypt a short stairway descends north-north-easterly to a little storeroom (serdab?) having six elongated chambers. These slots may have housed canopic vessels, shrines, or other funerary articles.

Menkaure's crypt

The most intriguing feature of Menkaure's pyramid, like Khufu's, is the burial chamber. Inside a room hollowed out of the rock the builders assembled a little house of red granite blocks. The walls and ceiling show craftsmanship rivaling that of Khufu's crypt. The ceiling is an arch of butting beams dressed on their undersides to the shape of a barrel vault.

Fig. 164 Menkaure's crypt in cross-section.

Fig. 165 Interior of G3 crypt.

The amazing thing about this ceiling is that it was put together in an extremely confined space. The peak of the ceiling is only about half a meter from the roof of the hollowed bedrock. That is not a lot of space in which to maneuver ten-ton blocks!

Fig. 166 The space above the ceiling beams of G3 crypt.

The builders may have pre-built this house outside of the pyramid. It is almost certain that they match-fitted the pairs of ceiling beams before taking them inside. They may have left levering notches on the undersides of these beams, as I suggested for the Queen's Chamber rafters in G1. The bosses would disappear in sculpting the curved ceiling.

But how did the builders put the ceiling blocks in place? It reminds me of the Sidney Harris cartoon where a scientist concludes his calculations with the expression "then a miracle occurs." The ceiling beams had to be inserted through a small opening in the rock at the south end of the crypt ceiling. The builders probably filled the room below with small blocks to support levers required for inching the granite beams sideways and to hold up one member of each pair while the other was gradually levered to its place. This construction is mind-blowing when we realize that the builders possessed neither screw-jack nor hydraulic ram. Even with modern equipment it would be difficult to duplicate this assembly.

As I said at the beginning of this chapter, Menkaure's pyramid is not lacking in quality of construction. Block for block it rivals the big boys, and might be the best built pyramid in Egypt.

The queen's pyramids

Three small pyramids, ostensibly for Menkaure's wives, lie close to the south side of G3. Reisner called them G3a, G3b, and G3c, east to west. Loose blocks of white Tura limestone are scattered about the southern base of G3a, suggesting it was completed as a regular pyramid. Not so for G3b and G3c. These appear similar to the stepped nucleus of big G3, but they couldn't have been finished as regular pyramids, as there is not enough space between them. The small chapels on their eastern sides also indicate that their stepped forms were meant to be final.

Fig. 167 G3a, b, and c, looking south.

Menkaure's pyramid temple

One final point of interest concerning Menkaure's pyramid concerns the mortuary temple that adjoins the east face of the pyramid. I mentioned earlier the large core blocks in Khafre's pyramid, mortuary, and valley temples. But Menkaure's mortuary temple goes one better. Here a 345-ton limestone block looks to have been used simply to fill a space between its walls!

Fig. 168 345-ton block inside G3 pyramid temple.

Why go to such an effort if the result was invisible and offered no structural integrity to the building beyond what an equivalent volume of small blocks (or even sand) could provide? What was the motivation for including huge blocks in hidden places? Was the value of a building related to the size of blocks it contained? Consider again the statement in Djehutihotep's tomb: "Behold, this statue, being a squared block on coming forth from the great mountain, was *more valuable than anything*." Perhaps the value of a stone accrued as much from the effort needed to obtain it than from its structural worth.

LATER PYRAMIDS

Compared to Dynasty 4 pyramids those of Dynasties 5 and 6 are not as impressive in size or execution. They are found at Saqqara and Abu Sir. Except for the pyramid of Neferirkara, all are smaller than Menkaure's. All have north face entrances and crypt ceilings similar to that of Khafre's pyramid. Their crypt ceilings have three layers of gabled beams. These pyramids, mostly in ruin, have nucleus structures in the form of large steps.

Neferirkara's Pyramid is interesting because it was apparently completed as a step pyramid without casing, and then enlarged in the manner of the Meidum E2 project. A casing of rose granite was started against the base of its first step, but never got further than the first course.

The Middle Kingdom pyramids (Dynasties 11 - 13) are even more dilapidated. Exposed by casing thieves, the less durable rubble and mud-brick interiors of these pyramids have suffered under the relentless actions of nature. A point for consideration in the later-discussed problem of material-raising: The bulk of these pyramids, consisting of mud-bricks and rubble, could have been manually hoisted right up the sides of the buildings.

The most interesting features of these pyramids are their internal layouts and robber-resistant crypts. A layout example is shown as figure 169, the Anonymous Pyramid of South Saqqara, Dynasty 13.

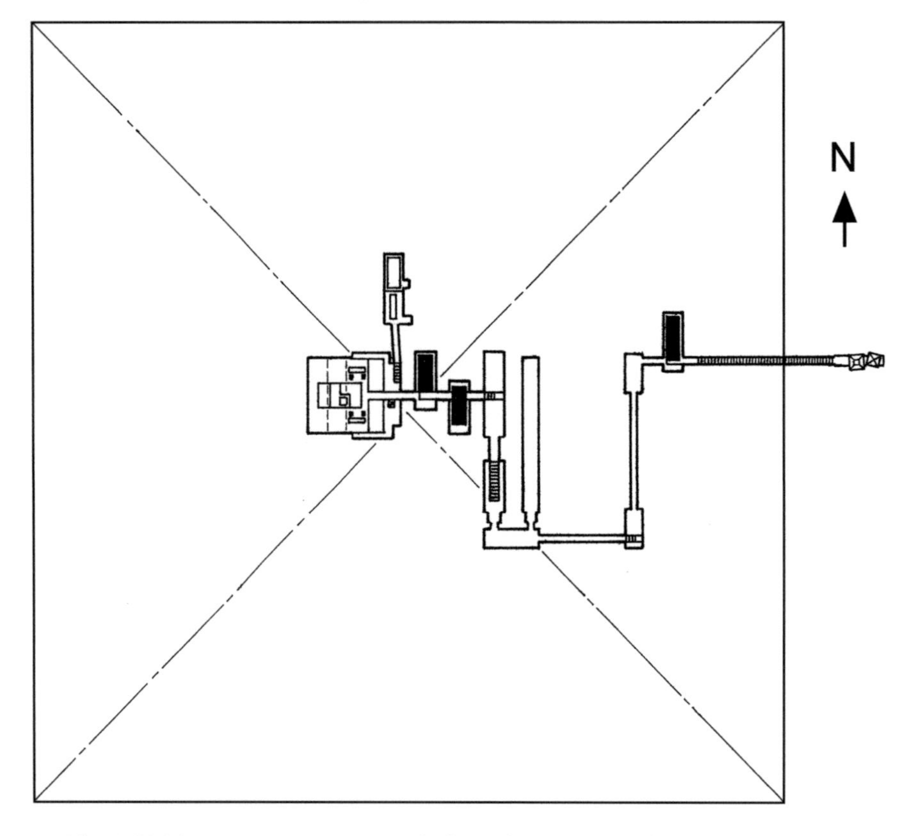

Fig. 169 The Anonymous Pyramid of South Saqqara. After Jequier (1933).

The Middle Kingdom architects must have noted the relative ease with which robbers penetrated earlier pyramids. They decided to make looting a little tougher. They no longer put entrances in the usual north face location. They made the route to the crypt more tortuous, and interrupted it with several portcullis gates of greater size.

The portcullis devices are large blocks set to move obliquely, but on such slight gradients that skillfully applied leverage would have been required to move them to blocking positions. Few, if any, of these gates were closed, so perhaps not much skill was needed after all.

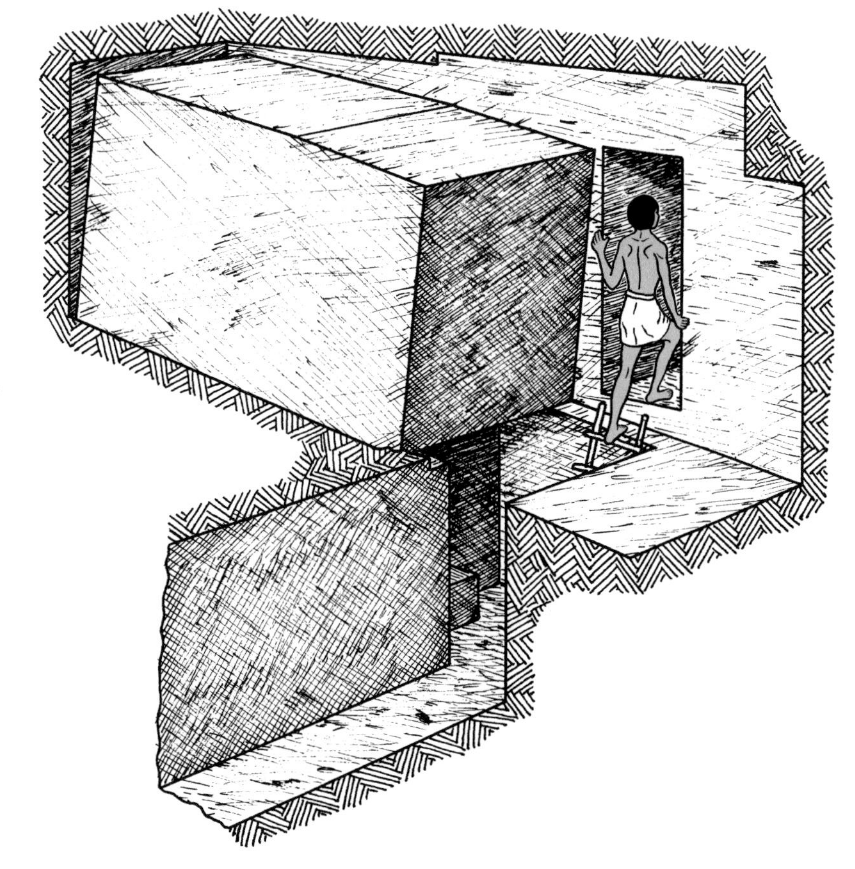

Fig. 170 Portcullis in South Pyramid of Mazghuneh. Adapted from Petrie, Wainwright, & Mackay (1912).

In Howard Hawks' movie *Land of the Pharaohs* (1955) a huge sarcophagus was sealed by slowly releasing sand from beneath stone pillars that supported the coffin lid. Amid rampant conjecture in the rest of the film, this portrayal is notable because it is based on fact. The principle was first used in the pyramid of Amenemhat III at Saqqara. The same construction is found in the pyramids of Amenemhat IV (Mazghuna), Khendjer (south Saqqara), and the Anonymous Pyramid of South Saqqara. The latter contains a 180-ton monolithic sarcophagus wrought from hard quartzite. The coffin has a three-piece lid, also of quartzite. One 78-ton segment is held up with temporary (waiting only four millennia) stone blocks, while the primary lowering means, a pair of quartzite pistons sitting on columns of sand, awaits actuation. The use of sand as a hydraulic fluid is another example of the builders' ingenuity.

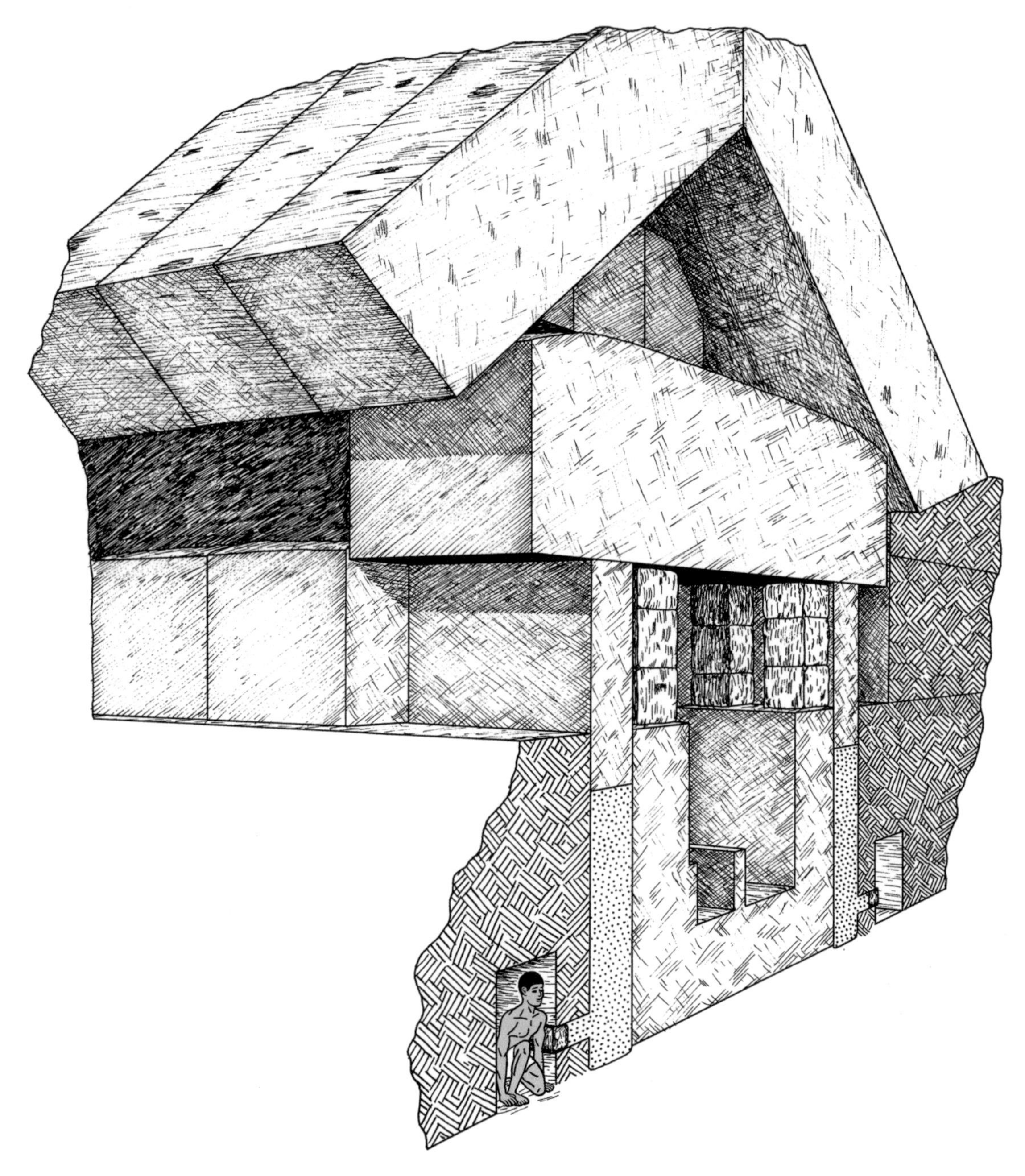

Fig. 171 Crypt of Anonymous Pyramid of South Saqqara. Adapted from Lauer (1976, 175).

9

ENGINEERING & MASONRY TECHNIQUES

Unless otherwise noted, the comments in this chapter apply to the Great Pyramid of Khufu. The Great Pyramid was ostensibly raised under the viziership of Khufu's cousin, Prince Hemiunu, owner of the beautifully-finished mastaba G4000 (Reisner) in the large cemetery west of the pyramid. I'll bet, however, that the design and planning of such a monumental work was a group effort.

Design

The first stage in producing the pyramid, perhaps concurrent with selecting its location, was to plan its design. This would have been done using line drawings on media such as papyrus, plaster, or limestone. Considering their affinity for models, the architects would likely have built one or more for study.

Layout and foundations

At the building site loose material was cleared, exposing bedrock on which the pyramid would rest. A rough survey established position and orientation of the building. The chosen location not only allowed a near-level perimeter, but also included a knoll that rose within the body of the pyramid.

A necessary operation in laying out a building is establishing its base perimeter. Here a perfect square of gigantic proportions had to be traced on leveled ground while maintaining the desired orientation (azimuth).

Because many elements - aligning, measuring, leveling - had to be simultaneously juggled, the surveyors probably did not fix the perimeter on their first try. Students believe that the surveyors must have approached the final alignment in successive stages of increasing precision.

The most comprehensive study of pyramid layout at Giza was made by Mark Lehner on behalf of the American Research Center in Egypt. As part of his work in surveying the Giza plateau, Lehner produced a detailed map of the shallow holes and other cuts in the pavement and rock around the two largest pyramids.

In *The Pyramid Tomb of Hetepheres and the Satellite Pyramid of Khufu* (1985, 54) Lehner says:

> Around the Khafre Pyramid there are two lines of small holes (30-40 cms. diam.) with roughly 10 cubit spacing, parallel to the base of the pyramid on all four sides. The most distinct of these lines occurs along the line of the outer edge of the enclosure wall, 13.5 ms. from the pyramid base, where holes occur in staggered pairs. Another line of holes, less regular than the first, occurs along the line of the inner side of the enclosure wall. At the four corners of the pyramid court, spanning the distance between the two lines of holes (and/or the width of the enclosure wall), are 6 m.-long trenches whose ends facing into the court are directly in line with the diagonals of the pyramid.

Lehner describes a similar series of holes spaced about 7 cubits (12 feet) apart around Khufu's pyramid. He believes that these holes "describe a great layout square around the pyramid base, both the square and base being determined with increasing accuracy over several runs by a method of successive approximation. As the layout square was refined, so was the baseline of the pyramid made more accurate by offset measurements."

The builders took exceptional care in placing the white limestone pavement that serves as a foundation for the first casing course of G1. The most important portions of the pavement were the corner slabs. Three were let into shallow sockets prepared in the bedrock. The fourth, at the southwest angle, was founded on built-up masonry.

The four sides of the pyramid vary from coincidence with the cardinal points (N, S, E, W with respect to the axis of Earth's rotation) as follows (Cole, 1925):

North side: minus 0° 2' 28" (89° 57' 32" true azimuth)
South side: minus 0° 1' 57" (89° 58' 03" true azimuth)
East side: minus 0° 5' 30" (359° 54' 30" true azimuth)
West side: minus 0° 2' 30" (359° 57' 30" true azimuth)

The greatest angular error, then, is nine one-hundredths of a degree, roughly 1/4000th of a circle. How can such an amazing alignment be explained? Many writers believe the surveyors could have achieved this degree of accuracy only by reference to the stars. Edwards suggested (1972, 200) a procedure similar to the following.

Two low walls are erected about 200 meters apart, east-west of each other, and equidistant from a sighting staff at point "C". The tops of the walls are made level with respect to the surface of the earth and each other, forming segments of an artificial horizon.

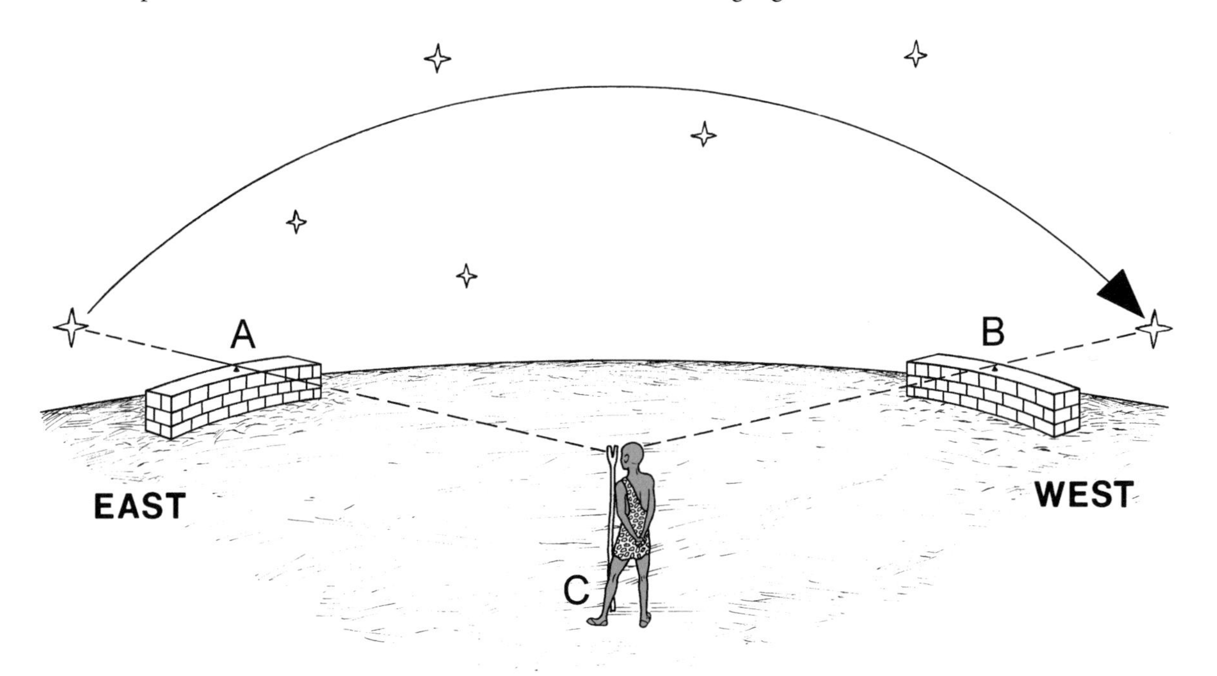

Fig. 172 Possible azimuth alignment method.

At dusk an observer sights, through a slot in the staff, the rising of a bright star as it appears over the eastern wall. He directs a helper to mark its position with a stone "A" on top of the wall. Near dawn of the following morning the same star is similarly marked with stone "B" on the west wall. Now a line connecting stones A and B will be in perfect east-west alignment.

For greatest accuracy the survey operation would have occurred during a winter month when darkness is longer, allowing more time for sighting the star. Sirius, the brightest star in the sky, would have been a good choice; in December 2500 BCE it rose in the southeast at about 5:50 pm and set in the southwest at about 4:14 am.

Let's say the north side was first aligned. Next would be marking off the east and west boundaries. For this a measuring technique, which I will propose shortly, was needed that would produce repeatable results for the other sides. The pyramid sides, at the perimeter of the original casing, measure (Cole):

North – 230.253 meters (755.42 feet)
South – 230.454 meters (756.08 feet)
East – 230.391 meters (755.88 feet)
West – 230.357 meters (755.76 feet)

The accuracy of figures given to the millimeter is suspect. Cole had no corner stones, not even corner pavement stones. All he had were sockets for the corner pavement slabs and, in places along the perimeter of the pyramid, chipped lines that had been formed in the process of final dressing the casing. We are stuck with his figures, however, because ablation of the pavement and sockets from thousands of shoes, feet, and hooves over the decades have rendered a more accurate survey impossible.

The longest side of the pyramid base is, then, only 201 millimeters (7.9 inches) longer than the shortest. The differences, though remarkably small, have been attributed to variable stretch in the measuring rope, a means Egyptian surveyors were known to use for land surveying. I wonder, though, if such a degree of precision could have been produced using something as resilient as a stretched fibre cord. Besides, the surveyors had a more accurate method available.

Measuring distances

Clarke & Engelbach noted that measuring *rods* can produce greater accuracy than stretched cords. To test this I took two eight-foot poles to a long, concrete culvert that ran past my daughter's school. Placing a 50-pound cube of granite as a starting point, I put the first pole against the cube and the second pole against the first. I leap-frogged the poles until I had measured ninety-six feet. I marked the ending on a card taped to the concrete. I made five more runs. The total range (shortest to longest) of my six measurements was slightly less than two millimeters.

Fig. 173 Use of measuring rods.

Scaling up to 230 meters this method would yield an error considerably less than that observed in the Great Pyramid, proving that this technique can produce a repeatable result. Since my experiment, Peter Hodges, in *How the Pyramids Were Built*, made the same proposal, suggesting the use of 20-foot poles. The use of longer poles would produce even greater accuracy.

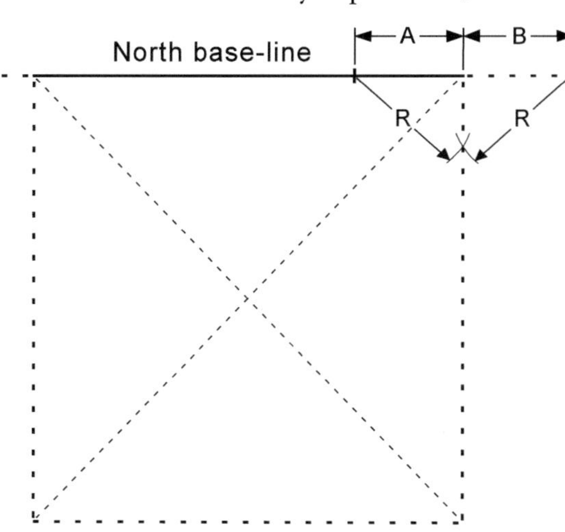

Fig. 174 Establishing the east and west sides of the pyramid.

Returning to pyramid base layout, after the northern boundary was fixed, the surveyors would have established right angles to direct the west and east sides southward. One way this could have been done is by using simple geometry as shown in figure 174.

The procedure would be:

1. Extend the layout line of the northern base-line to the east and west.

2. Mark off equal distances (A & B) from the corners of the pyramid.
3. Strike equal radii (R) from the ends of lines A & B.
4. A line from the intersection of the two radii to the corner of the pyramid will now form a right angle with the northern base-line.

At the required distance (440 cubits) from the north base-line the southwest and southeast corners would have been marked. Errors in square of the base could have been reduced by adjusting the length of the derived south side, if different from the north, and averaging repeated right-angle measurements from the northern corners. Square of the base could also have been checked by making sure the diagonals were equal. That procedure would have been challenging because the elevated bedrock core of the pyramid prevented leveled lines from corner to corner. However, a careful survey team could have used plumb lines in combination with their measuring rods so only horizontal components of the traversed distances were summed.

Establishing zero level

The pavement on which the first course of casing sits is flat and level within twenty-one millimeters (Cole) around the entire perimeter. A prevailing explanation for this astounding circumstance is that the builders used the surface of a water-filled moat as the reference datum. Lehner believes that the large, shallow depressions immediately outside the base perimeters of G1 and G2 served this purpose. He thinks the surveyors mortared short wooden posts into the holes and made the tops of these posts level with moats that encircled the pyramids. The post tops then became control points for further leveling operations.

This hypothesis is ingenious but unnecessarily complicated and messy. It's more likely the builders used a larger version of the A-shaped plumb level that we have seen in ancient drawings and one surviving example. A larger version would yield greater accuracy over longer distances. Also contravening the moat idea is that courses had to be level-checked (in

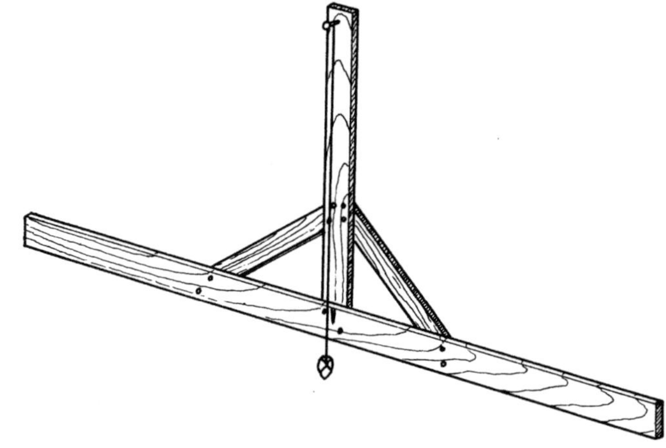

Fig. 175 The plumb level.

order to keep the building square) periodically as the building rose, and establishing a moat in the upper reaches was hardly an option.

Another experiment. I made an 8-foot-long plumb level as is shown in figure 175. A small stone hung by a length of sewing thread from a nail at the top of the vertical plank. The instrument was easy to calibrate, i.e. make capable of accurate measurements: I placed the ends of the horizontal plank on stakes which I had driven into the ground and made level by eyeball. Where the sewing thread crossed the horizontal plank I penciled a thin vertical mark. Turning the device end-for-end I again marked the thread position. A third mark, bisecting the space between the first two, would now indicate true level.

On a windless day I took the level and a bundle of stakes to a large open field. I hammered stake #1 into the ground and then leveled stake #2 with #1. I leveled stake #3 with #2 and so on. I repeated this process twelve times, making a large circle, 96 feet in circumference, back to stake #1. Each time I reversed the level to reduce errors of calibration and reading. The last stake placed beside stake #1 differed in height by a scant 1.5 millimeters. This error would equate to 23.6 millimeters for half the perimeter of the Pyramid, the greatest distance from the starting point in any direction toward the far corner. This accuracy is only slightly less than the twenty-one millimeters found by Cole. With a larger version of this level, and averaging several tries, I believe I could equal or surpass the accuracy seen in G1. Although this experiment does not disprove the moat theory, it shows the availability of a more direct and simple procedure consistent with known building methods.

As for the shallow depressions surrounding G1 and G2, I prefer the explanation of Maragioglio & Rinaldi: These channels were rain gutters that allowed water runoff that otherwise would collect in unsightly pools around the buildings.

Skew, Slope, and Hip Lines

After the builders had established the square and level base it was time to raise the pyramid. To build accurately, several factors had to be constantly checked. These are skew, slope, level, right angles at corners, length of sides, and straightness of the hip lines.

Skew is twist of the pyramid edges, as shown in an overhead view in figure 176. The prevention of skew is fairly simple. As Clarke & Engelbach noted, a person standing well outside the pyramid base, and in line with opposite corners, could have aligned the edges by sighting along a plumb line fixed for the duration of construction.

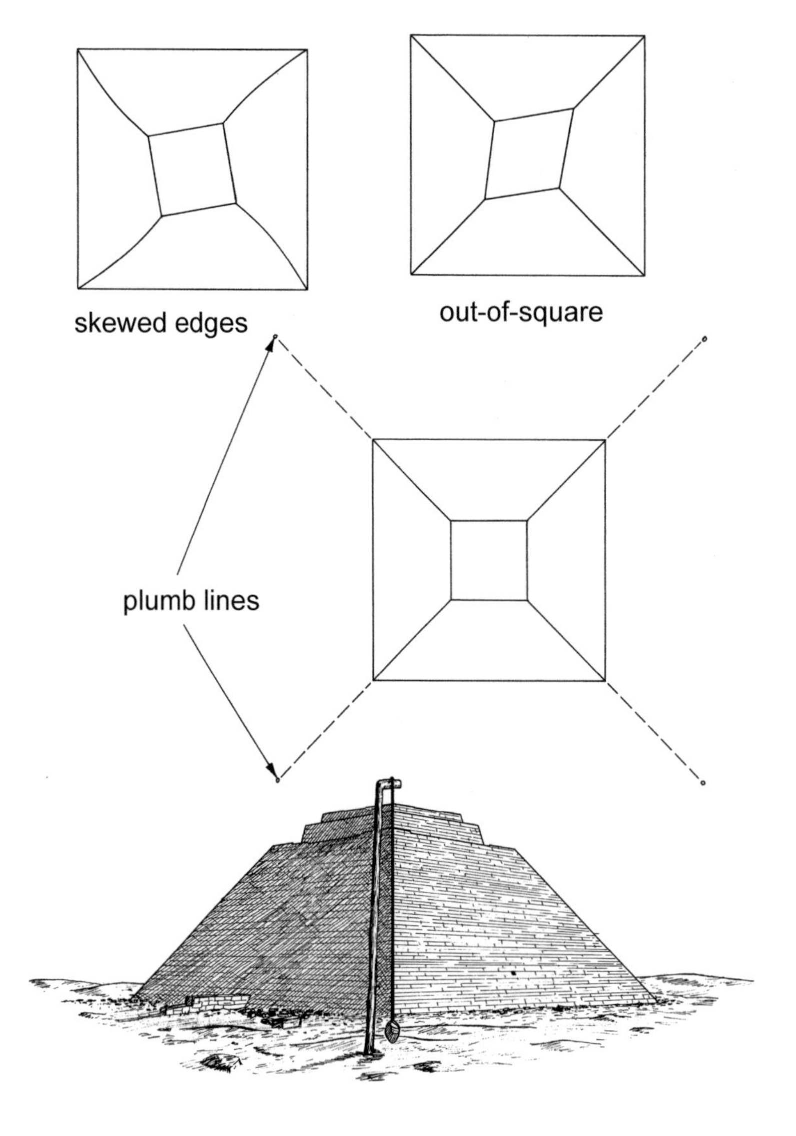

Fig. 176 Edge alignment using a plumb line.

Side slope was probably controlled by several techniques. Slope gauges of various sizes could have been used to check slope from course-to-course or over several courses. Though the configuration shown in figure 177 is hypothetical, a plumb-checking device of similar construction can be seen in the Egyptian Museum.

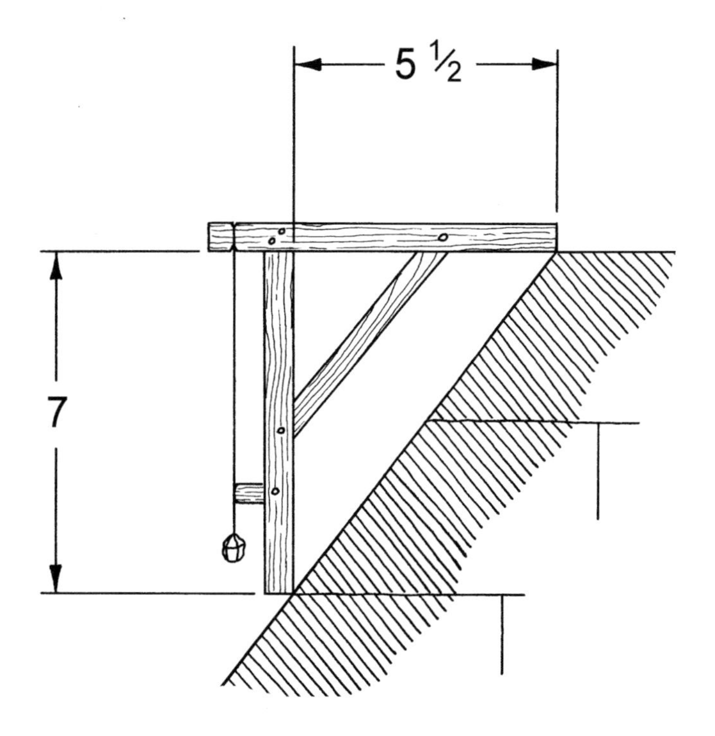

Fig. 177 The slope gauge (hypothetical) for G1's seked of 5 ½.

Slope accuracy was probably checked periodically by measuring the square at the top surface and comparing it to the calculated value for that height above zero level. Large errors in slope could have produced the out-of-square condition shown in figure 176. Keeping the diagonals equal would have prevented this condition, but since the inner core structure may have been in the way it may have been necessary to make the *core* diagonals equal and measure offsets to the edges of the outer (true pyramid) form.

Now we come to a pyramid characteristic that must have been, for both aesthetic and practical reasons, most important to the builders: straightness of the hip lines. Hip lines are those formed by the intersection of faces of the pyramid with each other. But where "corner," "edge," or "angle" could refer to the general vicinity of the corners of the pyramid, "hip lines" refer to the actual sharp edges of the intersections. Look again at figure 144, the top of Khafre's pyramid. Notice that the hip lines are not only straight, but if extended would meet at a point, an obvious requirement. One can also observe straight hip lines on the casing at the northeast angle of the Bent Pyramid.

Fig. 178 Straight hip lines on NE angle of the Bent Pyramid.

How can such outstanding alignments be produced? Mendelssohn suggested (1974) that a step pyramid was first constructed. On its top a pole was erected, the top of the pole then being sighted along the edges of the true pyramid as the steps were filled. Thus, skew prevention, slope accuracy, and dead straight hip lines were all maintained by observing the same datum. Evidence he provides in support, however, is weak. He cites an observation by M.A. Robert, mentioned earlier, who noticed on the top center of the Meidum pyramid a hole, 15 cm (5.9 inches) diameter by 30 cm (11.8 inches) deep. Robert surmised that this hole "served as a receptacle for a mast of some kind."

If this hole held a mast it is unlikely that the mast was a tall one, even with supporting guy lines to keep its tip from wandering in the breeze. The hole more likely held a short pole, possibly serving as a datum for measuring distances to the pyramid edges.

On the Bent Pyramid, if Mendelssohn is right that a stepped nucleus was built entirely before the true pyramid shape was begun, it would have been necessary to remove a large number of blocks from the stepped upper section of the nucleus when the pyramid angle was reduced. To me, that seems unlikely because the builders would have had to admit to a huge waste of effort. Conversely, if the (hypothetical) stepped nucleus had not been advanced beyond the bend level, a pole anchored at that height would have been over eighty meters (263 feet) tall!

Further, I would like to mention that there is no central hole on the top level of the Great Pyramid.

Perhaps hip lines were kept straight by building to a stretched cord, as stone masons do today. To support cord-holding posts is a good reason, among others, why stepped cores were built and advanced somewhat ahead of the casing and filling for the final, regular pyramid form.

Fig. 179 Building to the hip lines delineated by a stretched cord.

Guiding hip lines in this manner, along with measuring a leveled square every few courses, and sighting the hip lines from points well outside the pyramid as shown in figure 176, could have produced the accuracy we see. This idea would carry more weight if the corner casing blocks on Menkaure's pyramid had remained *in situ,* as these stones might show trimmed hip lines consistent with this proposal. Sadly, the edge blocks are missing. Perhaps some of these are among the granite casing stones now scattered around the pyramid.

Masonry rules

From observing many buildings, certain rules of Egyptian masonry practice become apparent:

1. The outer faces of blocks in a vertical *wall* remain in the rough until a certain expanse of masonry can be finish-dressed. Then the dressing proceeds from bottom to top. This rule, however, may not apply to pyramid faces, where complete smoothing of the lower courses would not allow traversing by the builders. Thus the dressed portions of granite casing we see on Menkaure's pyramid (described later) may be atypical.
2. Vertical joints are always broken (never align from one course to the next).
3. In pyramids, each course of casing is the same height (thickness) all around the perimeter. In vertical walls, casing stones are not kept in distinct courses.
4. A block having a fitting cut was placed before the one that was inserted into it.
5. Where possible, casing stones are placed first, followed by backing stones, then core stones.
6. On pyramid courses, corner stones were placed first, followed by blocks proceeding toward the middle of each face.

Fitting cuts

Rule "4" deserves more attention. I have mentioned fitting cuts before, but what is their significance? Egyptians constructed masonry walls using stones that varied considerably in size and shape. This permitted quarry gangs to produce blocks sized according to quarry conditions (seams, joints, strata spacing), and the handling capacity of gangs. Not wanting to squander material arduously quarried and transported to the building site, masons cut to waste the least amount. Intermixing blocks of unequal height would then produce the situation shown in figure 180, where corner "E" would interfere with the next block placed.

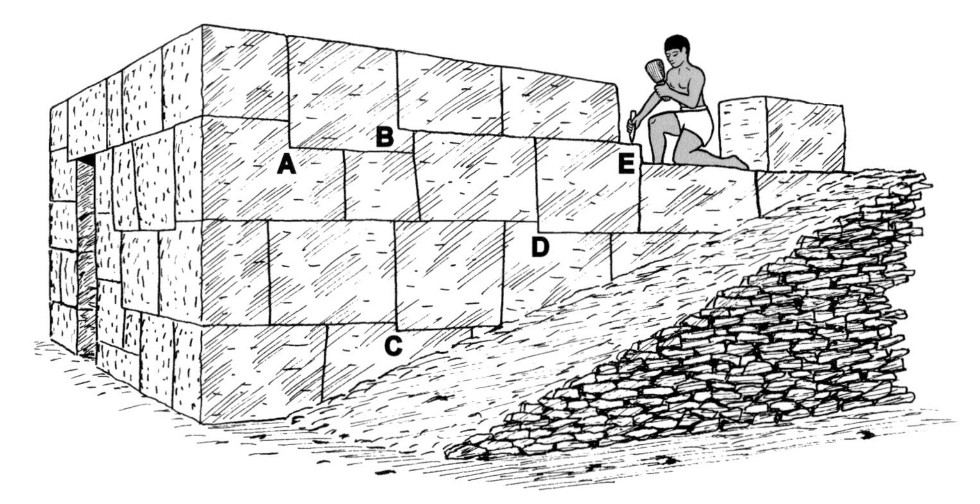

Fig. 180 Fitting cuts.

What to do? It was much easier to cut corner E away than to make a clearance for corner E in the underside of the next block above. Fewer faces had to be match-dressed. The cut-away corner at E is a fitting cut, as are those at A, B, C and D.

In figure 181, block A has a fitting cut in its upper right corner where block B was inserted into it. Notice also in this photo the upper blocks that have not been finish dressed.

Fig. 181 Fitting cuts in walls of mastaba in the Western Field Cemetery.

Fitting cuts are always in the upper surfaces of blocks, never the lower. Since we can agree (I hope) that walls are constructed from the bottom up, it logically follows that *fitting cuts show the order of placement of blocks.*

The fitting-cut rule gains significance when we realize it is also applied to horizontal courses of pyramid masonry. With respect to the top course of blocks on G1 (figure 160) I said that the fitting cut in block A shows that block B was laid after it. That fact, plus the narrow slivers of stones near the south border of this course, suggests that the interior was filled from north to south. This may indicate that the blocks for this course arrived at the south side.

Builders' marks

The builders painted reference lines in red ochre on masonry walls to aid their measurements, particularly in leveling operations (figure 159). On these lines they drew triangles pointing in directions from which measurements were taken. The value of the measurement is usually beside the triangle. On the south wall of Khufu's eastern boat pit an unusual variation appears in a line of ten symbols. These incorporate the cubit sign (lower arm of human) as the horizontal leg of the triangle, the measured value being within the triangle itself.

The builders' marks include colorful working-gang names. I mentioned two of these in chapter one. Others collected by Alan Rowe, who worked with Reisner, are:

The gang - Neferka is friendly
The gang - Khufu excites love
The gang - The white crown of Khufu is powerful
The gang - Sahuru is beloved

My favorite inscription was found by Petrie on two blocks that had fallen from the ruined Meidum pyramid. It translates "this side up"!

Block incisions

In the blocks of the Giza pyramids, especially the two largest, one can observe occasional holes and notches the purposes of which defy easy explanation.

I group the incised blocks into several types:

Type A – Rectangular holes in tops of blocks.

Fig. 182 A pair of type A holes on top of G1, SE corner.

Five examples in the backing stones on top of G1 (marked locations "C" in figure 160) are about 10x20 cm x 6 cm deep. The similarity of orientation suggests that 1) they were cut in already-placed blocks and 2) they had a functional connection. Two other examples a couple of courses lower, on the north side, are of similar size. Both of the latter are beveled on their outer (northern) sides. Hypothetically, these two holes could have held the ends of wooden poles that were be rotated, in one direction or the other, between vertical and an angle slanting away from the pyramid face – for what purpose I do not know.

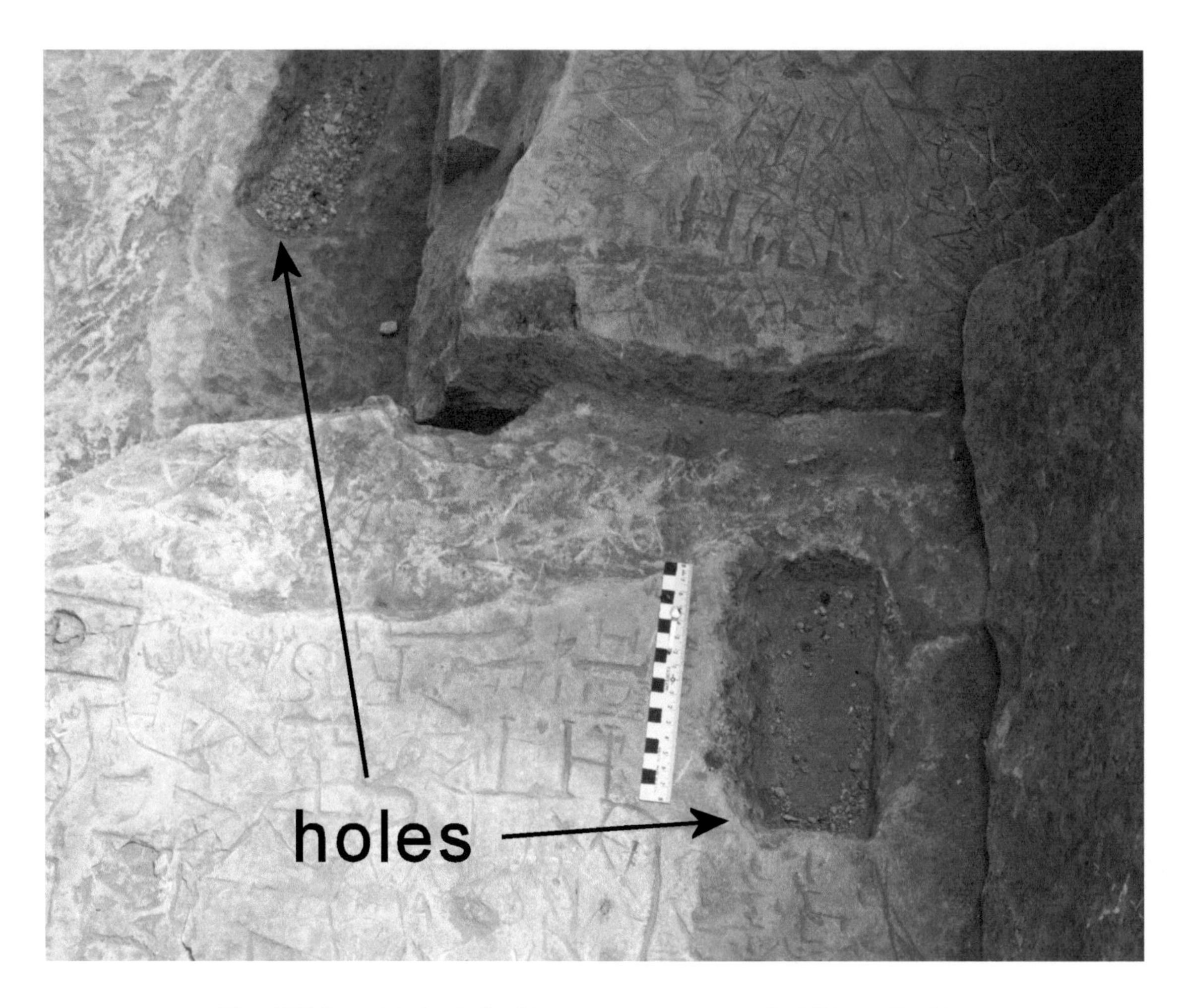

Fig. 183 Type A holes in backing stones near summit of G1, north side.

The better-preserved examples of type A holes have chisel tracks in their bottoms. This would indicate that they either held static objects or were partially filled with a material (perhaps sand) that prevented abrasion by the objects inserted into them. Type A holes could have served as handspike fulcra, or perhaps held poles from slipping sideways.

Type B – Square holes in tops of blocks.

More rare than type A, these may have held wooden posts used in leveling or other measurements.

Type C – Round holes in tops of blocks.

Fig. 184 Top of G1, blackened hole, 17 cm diameter.

These depressions are usually about 10-15 centimeters in diameter. Two examples on top of G1 (marked "D" in figure 160) have hemispherical bottoms. The one to the east has slightly conical sides coated with a jet-black substance that appears carbonaceous. Its bottom looks rusty. These holes may have served as bearings for rotating poles used in the manner of a capstan, though this is strictly conjecture.

Type D – Notches in lower edges of blocks.

Beveled crescents and rectangular bevels are probably levering points used either for breaking blocks from their beds in the quarry or for making small adjustments of position on the pyramid.

Fig. 185 Type D notches in first course backing stone, east side of G1, north end. The left notch is mortar-filled.

Type E – Holes in sides of blocks.

Type E holes may have been used to momentarily lift a stone to adjust its position or remove it from its sled. Some may be measuring points, especially those in a first course backing stone on the north side of G1, about 50 m (164 feet) from the NE angle. This block, seen in figure 186, contains two type E holes near its upper edge. It and its neighbor to the east (left) also contain fitting cuts showing that the casing blocks now missing were laid from east to west. The narrow slab to the right of this block may be the closure block for this row of backing stones.

Fig. 186 Type E holes in first course backing stone on north side of G1.

Type F – Rectangular notches in upper edges of blocks.

These notches would have allowed levers to lift the stones in maneuvering or rolling them if they were originally at the lower edges of the blocks, i.e., first serving the same as type D notches.

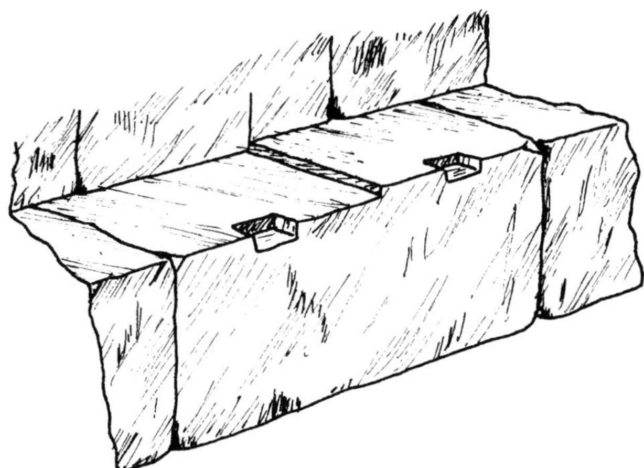

Fig.187 Type F holes in backing stone on east side of G2, south end, second course.

I emphasize the infrequent occurrence of blocks having cuts, notches, or holes. Less than one percent of visible stones have any sort of cut. That makes the incised blocks exceptional and not the rule. Also note the *absence* of rope notches such as found on the blocks covering Khufu's boat pits, which I will cover soon. For now we can conclude that the builders didn't need to make notches or holes in their normal handling of the masonry.

Handling the blocks

The builders moved stones about with the simplest of tools.

Foremost in the block-mover's tool kit had to be the ubiquitous lever, one of six simple machines found in devices of greater complexity. The others are the ramp, screw, pulley, wedge, and wheel-and-axle. All can be used to gain *mechanical advantage* in moving a heavy load a short distance by applying a lesser force over a greater distance. The Egyptians were undoubtedly without peer in masterful use of leverage.

Because no anciently-dated levers of metal have been found we must assume that wooden levers were employed. Besides, the large notches found in some blocks indicate rather large cross-sections of the points used, consistent with wooden levers.

To gain the highest mechanical advantage and minimize bending stress in the lever, the pivot point, or *fulcrum,* must be placed as close to the load as possible.

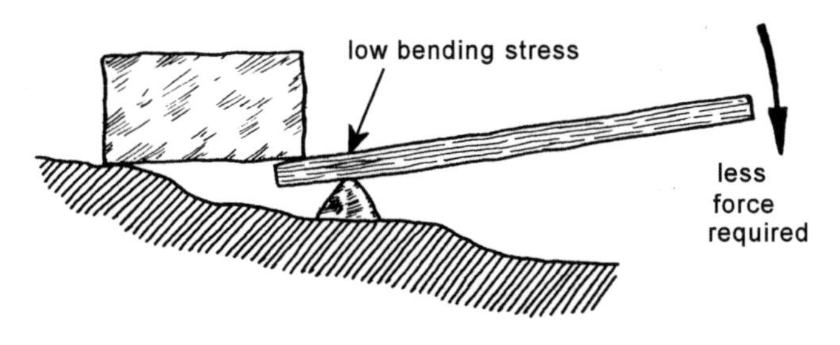

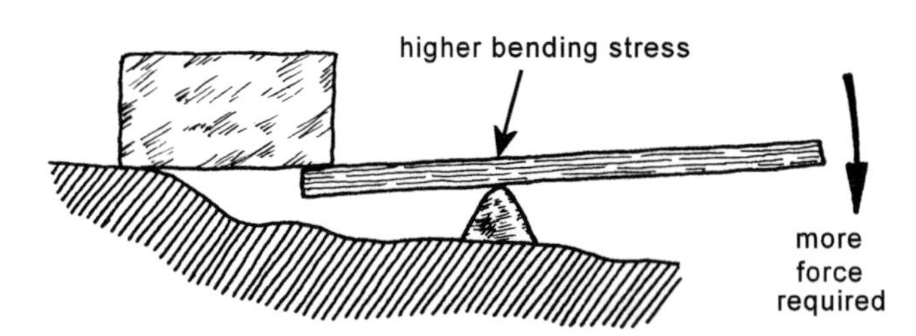

Fig. 188 Fulcra placement versus bending stress.

A series of levers can be employed not only to lift a block but to move it horizontally in short increments.

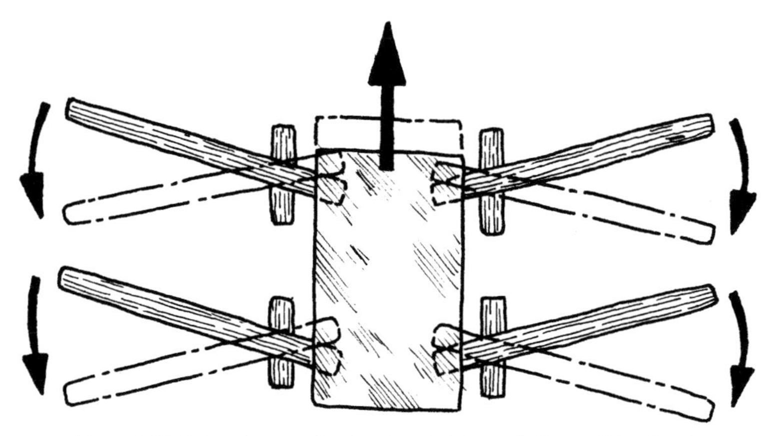

Fig. 189 Using levers for horizontal translation of the load.

A lever can also be used to push an object along the ground using an anchor point in the ground as the fulcrum. A lever used in this manner could be called a *handspike*.

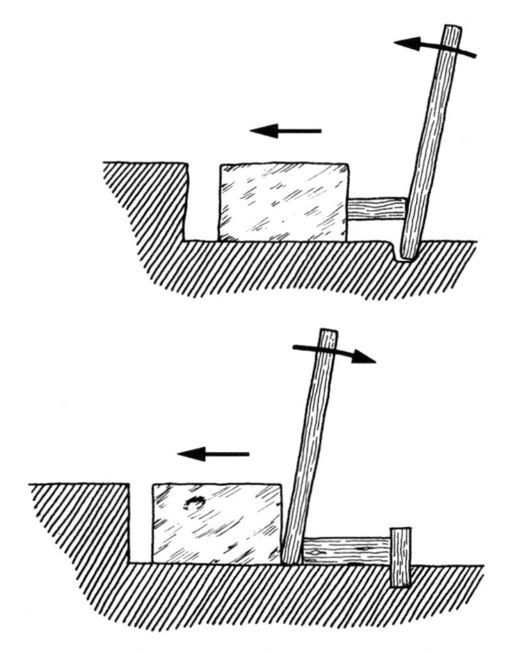

Fig. 190 Lever fulcrum in ground.

Levers aren't restricted to direct action against an object. They can be employed along with ropes to lift or pull an object.

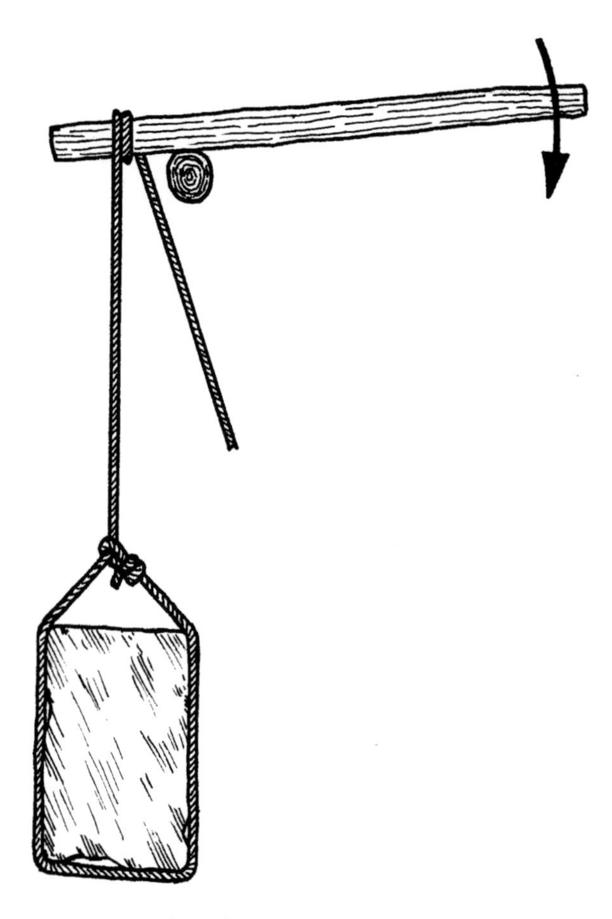

Fig. 191 Lever-and-rope use.

A simple technique for rotating a block on a hard surface is to lift one end slightly and place a small, hard stone under it near its center. When the block is lowered onto this "pivot" it can be easily turned to a new direction.

To lift or push really large stones the workers may have attached ropes to long levers to allow manipulation by many. It was by such means that the Assyrians helped move the colossal winged-bull statue of King Sennacherib, as depicted in the relief sculpture in his palace at Nineveh (c. 690 BCE.).

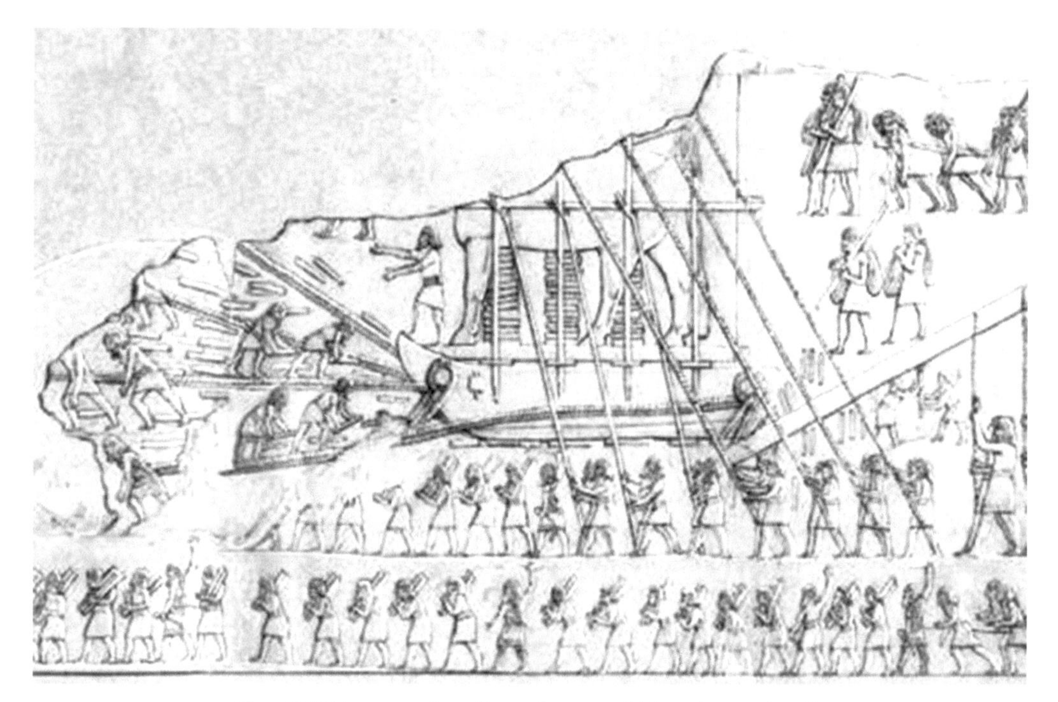

Fig. 192 Transporting the colossus of King Sennacherib.

Laying the core blocks

The core blocks laid on edge in the upper part of Khafre's pyramid were almost certainly rolled into position. These blocks are twice as high as they are wide, and would have been transported, I think, in a more stable (flat) orientation, then rolled into position. I wonder if this might have been the procedure for most or all of the core blocks.

How does one roll a multi-ton block? Since most of the pyramid stones lack lever notches it would have been difficult to insert levers under a block sitting on a hard, flat surface. On the other hand, it would have been fairly easy to roll or lift the block if it was sitting on a *pallet* of some kind. But what kind of pallet? Consider the transport sled described in chapter three. It makes a perfect pallet.

Because a block would normally overhang the sides of a sled, levers could be inserted under one or both sides of the block, depending on whether it was to be rolled or lifted. The block could also be rolled in a forward direction using the sled itself as a double lever. In this instance other levers could be put under the beveled chamfers at the rear underside of the sled to start the lifting process. The upturned runners at the front of the sled would have assisted the rolling process by preventing the block from moving forward prematurely.

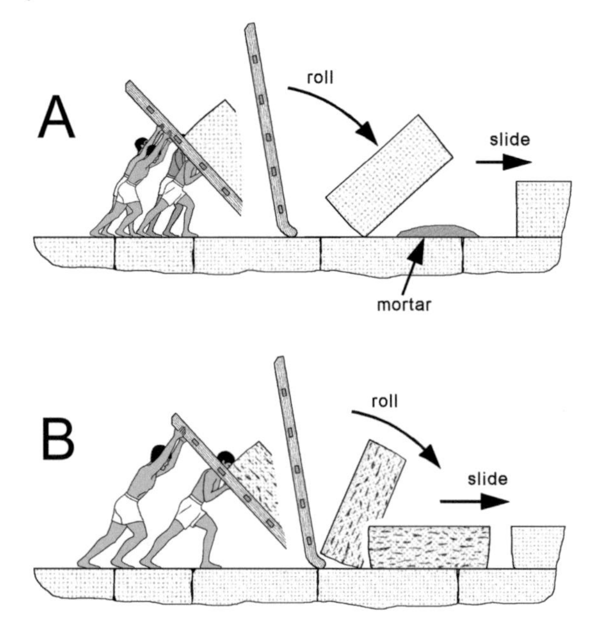

Fig.193 Rolling a block using the transport sled with mortar as a cushion and lubricant (A), or no mortar (B).

From experiments I have done with small blocks of limestone, the block, when rolled in this fashion, will jump forward upon landing, and even with no lubricating mortar it will slide up to one quarter of its length, bringing it flush against the previous block laid. The transport sled, then, may have been a multi-purpose device, a tool for block maneuvering and placement in addition to transport.

Laying the casing stones

Clarke & Engelbach believed that pyramid casing stones were always laid from the front, or outside, of the pyramid. The procedure they envisioned (1930, 110) required a "platform of brick and rubble extending some forty feet out from the stonework."

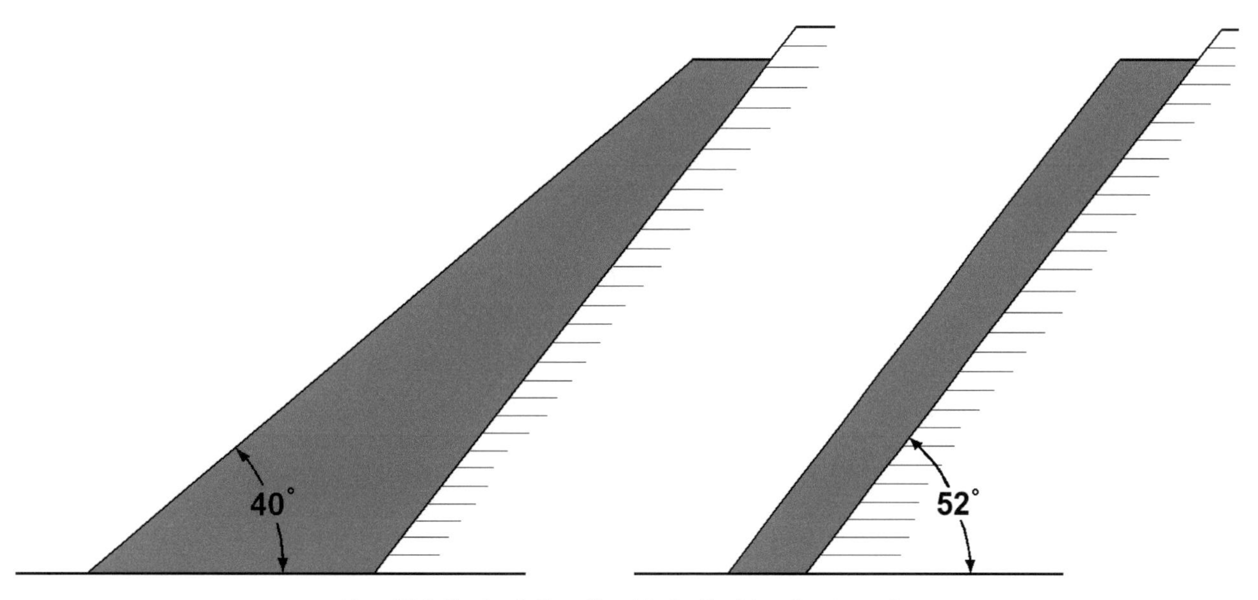

Fig. 194 Clarke & Engelbach's foothold embankments.

But these embankments, going all the way to the ground, would have required a huge amount of material. If their faces sloped at the same angle of the pyramid, fifty-two degrees, their total volume would have been a third the volume of the pyramid; if they sloped at a safer angle of forty degrees, their volume would have been almost twice the volume of the pyramid! And these figures do not include the volume of the ramp over which the stones were raised to the top of the embankments.

I think it's more likely that most of the casing stones were pushed or rolled to their final positions *sideways*, one against the next. Evidence can be seen as lever notches on the sides of casing stones of Khufu's pyramid.

Fig. 195 lever notches in the side of a casing block on west side of G1.

Other evidence against the outside-to-inside direction of placement might be a protrusion on an 18-ton granite casing block on the west side of G3. The protrusion is higher than its bedding surface for the course above.

Fig. 196 18-ton granite casing stone on west side of G3.
Greg's left hand rests at the base of the vertical protrusion.

The casing stones on the north side of Khufu's pyramid may be the best-fitted megalithic blocks on Earth. How these huge stones could have been brought into such close contact is not easy to fathom. The most difficult task for the masons was match-fitting a first-course casing block simultaneously to its neighbor and its own backing stone. This situation arose when proximity of elevated bedrock either forced placement of the backing stone first or required placement of the casing stone directly against the bedrock. In these cases, two rising joints had to be matched instead of one. I should mention, however, that the tight fitting between the backing stones and casing at the north-central group on G1 may be atypical. The backing stones on either side of this north-central group are not nearly as tight. So too, the vertical joints between backing stones become much wider as the courses rise.

Fig. 197 The loose fit-up of backing stones on north face of G1.

Still, we must explain how the tight joints in the north-central row were accomplished. The following explanation of this process is my best guess.

The near-vertical faces (some were oblique) of the blocks already placed had been dressed flat beforehand. The surfaces were checked first by using straight-edged planks and boning rods. Boning rods are a set of three pins, two of which are connected with a stretched cord. The cord is fastened to the two end pins at a height equal to that of the third pin. When the third pin is moved back and forth under the cord it indicates high or low spots on the surface being checked. Flat, ochre-coated boards, might also have been employed. The ochre-coated plate, when rubbed against a surface, transfers paint to high spots of the surface, again showing where material should be removed.

The new block, with its bottom surface dressed flat, is maneuvered, still on its transport sled, to the proximity of its laying bed. Its outer face, except perhaps for a small ridge of extra stock at its lower outside edge, has been dressed to the pyramid slope.

Again comes the idea I referred to earlier. The new block is resting *on its side* on its transport sled because it is going to be rolled into position (figure 193-A). I have long thought that most of the *core* blocks were rolled into position, but doubted the *casing* stones were so-placed because of the danger of chipping their edges. On the other hand, I had trouble envisioning how the casing blocks, especially the massive ones in the first course of G1, were levered into position without having wood scrapings from the levers remaining under the blocks. Wood fragments would have prevented flush seating of the bedding joints. Further, I wondered how a multi-ton casing block could end up tight against its two neighbors.

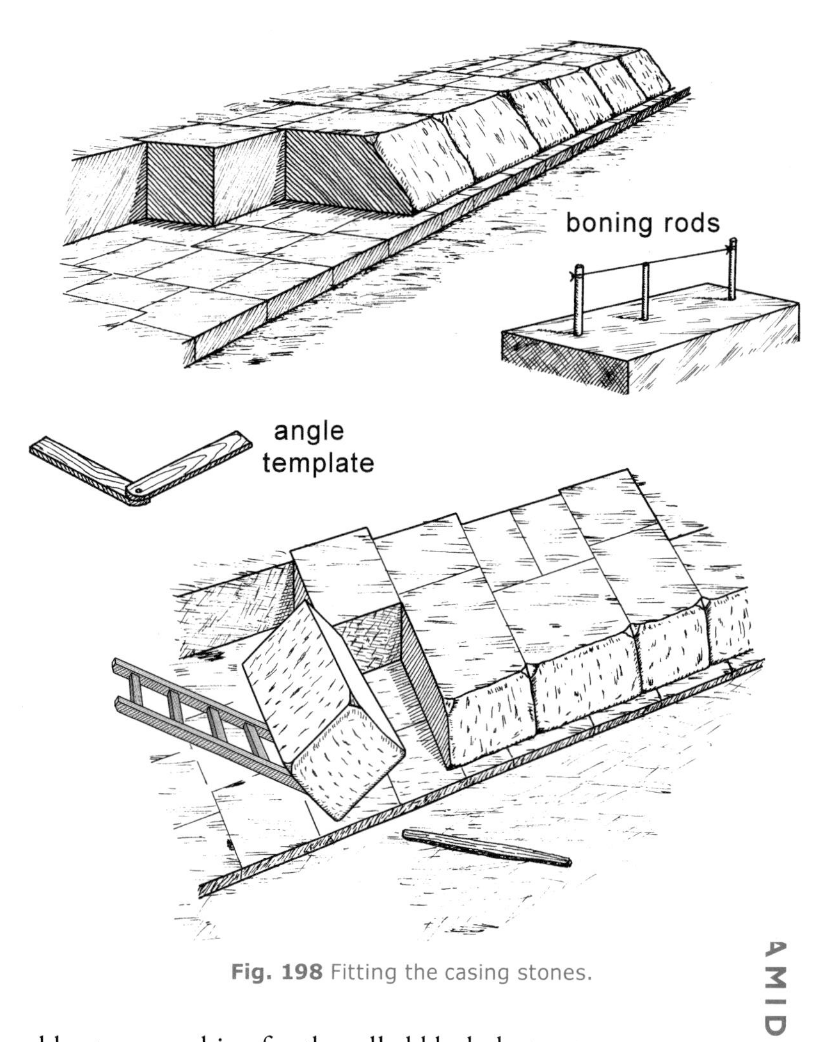

Fig. 198 Fitting the casing stones.

Another experiment. On my garage floor I placed a fifty-pound cube of granite on my model sled. I mixed a blob of plaster-of-Paris to a consistency that I thought would act as a cushion for the rolled block, but not so viscous that it would prevent the block from fully seating against its neighbor. My garage door would represent the neighboring block.

Upon tipping the sled up, the block rolled onto the plaster-cushioned floor. Splat! It threw plaster in all directions. Having forward momentum, the block slid up against the garage door. However, I could slide the block around like it was on ice. This proved that the procedure was at least feasible, and that Clarke & Engelbach's proposal that gypsum plaster was used mostly as a lubricant is probably right.

I should point out here that I didn't observe traces of mortar on the bedding surfaces of blocks higher on G1than the first course. I see coarse mortar between the blocks, but not on the bedding surfaces. The bedding surfaces reveal tool cavities unfilled with mortar.

Fig. 199 Bedding surfaces on corner blocks about three quarters up the southwest angle of G1.

Back to the casing of the first course. While the casing block was still on its sled, and before being tipped on its side, the masons could have match-dressed its faces to the faces where it would seat by using angle templates to transfer the angles of the seat faces to those of the block on the sled. The angle template shown in figure 198 is a gauge made of two wooden planks pinned together tightly enough such that when one is rotated with respect to the other it will retain the angle at which it is set.

But I don't think the use of angle templates alone can account for the accurate fit-up between these casing stones.

Placing the casing stones, especially the larger ones, was probably a one-shot deal. The stones had to fit correctly on the first try because the mortar begins to harden in about 15-20 minutes if properly prepared. Baking calcium sulfate hydrate (CaSO4•2H2O) drives out part of the water such that when water is reintroduced to the powder, the plaster will recrystallize (harden). Plaster hardened or not, there was no way to remove an ill-fitting block without damaging it. Proof of this can be seen on Khafre's pyramid, where medieval quarrymen found it easier to pilfer first-course granite casing stones by wedge-splitting the outer portions away rather than trying to remove them intact. This makes us appreciate the precision of the ancient masons all the more.

American sculptor and pyramidologist Martin Isler (1926- present), in *Sticks, Stones, and Shadows, Building the Egyptian Pyramids* (2001), suggested that final match-dressing was accomplished in manner similar to that found by Gantenbrink for the air ducts in G1, that is, by running a saw blade between blocks that would adjoin. But Isler's new idea was that the match-dressing-with-saw-blade was done between the block about to be placed *and the next one following*, not between the block about to be placed and the one already in position on the pyramid. This is an excellent guess.

As mentioned earlier, the need for accurate match-dressing of the casing blocks was a priority only for joints visible when the pyramid was finished. Clarke & Engelbach said "In many of the Fourth and Fifth-dynasty mastabas at Giza, masonry can be observed where the blocks fit tightly at the face but the joint only extends inwards for at most a couple of inches." Thus, the interior of the block faces could be slightly hollowed so as to not interfere with each other. In other words, it was important to err on the side of non-interference in the middle of block faces.

For most casing stones, those in upper courses where workmen could get to the rear sides of the blocks, the masons may have employed the saw for trimming-to-fit between blocks about to be placed and those already in place, as shown in figure 200.

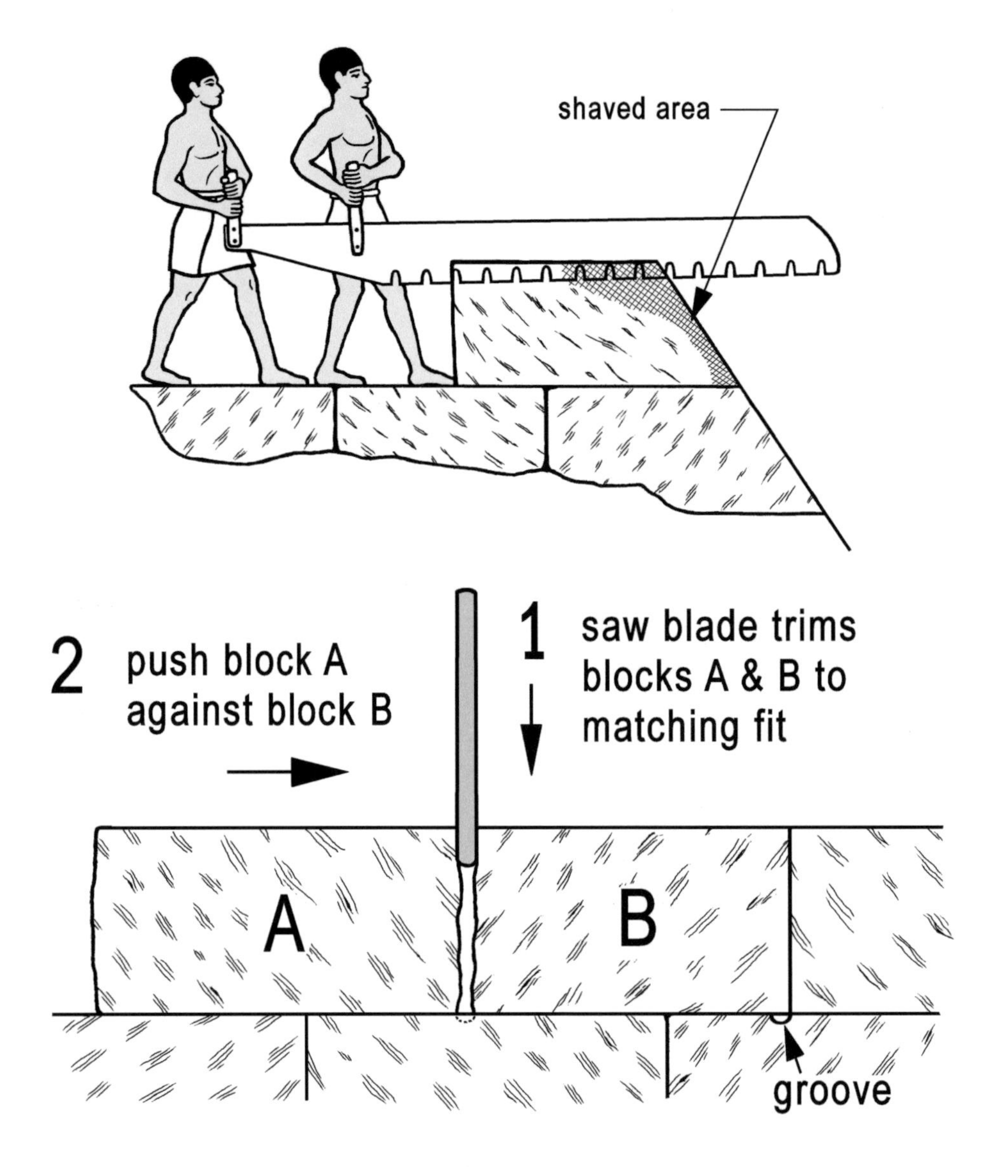

Fig. 200 Match-fitting the casing stones.

In 1987, additional evidence could be seen on casing blocks of G2 that had fallen to the west base of the pyramid. On the block shown in figure 201 you can see that its side is smoothest at the outermost portions of the joint. This condition is consistent with the technique just described.

Fig. 201 Fallen casing stone on west side of G2.
The scale is on what would have been the outer (sloped) face of the pyramid. Note the step in what would be the bedding joint for the next higher course.

There is another aspect of the casing block in figure 201 that is remarkable. Notice the exquisite flatness of the outer face (on which the scale rests) and the top surface (in shadow). This block must have fallen recently, and thus would have been part of the casing at a height about three quarters of the way up the pyramid. It is noteworthy that the masons did not try to take liberties where the powers-that-be might not notice. It looks like they did high quality work for its own sake.

Clarke & Engelbach surmised that stretches of casing were match-dressed in a building yard, then placed on the pyramid with no additional fitting. This seems improbable. Fit-up conditions on the pyramid would not be close enough to those in the building yard to consistently produce such tight joints. In many backing stones on Khufu's pyramid one can see individually tailored bedding seats for adjacent casing blocks, seats that do not align. The slight step in the middle of the top of the casing block shown in figure 201 indicates that the bedding joint for the next higher block was cut into this block after it had been placed on the pyramid. Thus, it is probable that each stone was fine-tuned close to its final position on the pyramid.

It appears the casing dressers did not require scaffolding on the pyramid faces. Captain Caviglia reported patched-over "putlock" holes in the casing of the Meidum pyramid and the pyramid of Senusret III. He surmised that these holes had received scaffolding props.

When I visited the Meidum pyramid I forgot to look for these holes. However, none are visible in my photographs of the casing. My telephoto shots of the top of Khafre's pyramid do not show any patched holes, nor did I find any in fifteen damaged casing stones that had fallen to the west of the pyramid. We must conclude, therefore, that these holes appear too infrequently to have served as scaffolding anchors. Perhaps Caviglia saw only repair work on flawed stones, a cosmetic improvement often seen in Egyptian masonry.

Finish-dressing the casing

An unresolved question: What did the casing stones look like when they were set on the pyramid? Were they left as stair-steps to be dressed down in the final building stage? This aspect would have allowed footholds for masons as they trimmed blocks from top to bottom. However, there is no evidence that this was the case. There *is* evidence as follows:

1. On Menkaure's pyramid, smoothing of its granite casing was begun at the bottom center of each face. At lower levels it is likely that ramps or embankments were laid against the pyramid faces, so it was not difficult to get to the casing for final dressing.

2. According to Goneim (1956, 47), at Sekhemkhet's pyramid the five lowest courses of the unfinished wall surrounding the complex had been finish-dressed, "the sixth and uppermost course had not been faced and left rough."
3. In several mastaba tombs in the Western Cemetery of the Khufu complex one can see walls that were being dressed from the bottom upward. The highest blocks have not been dressed flat (see the blocks indicated by arrow in figure 181).

If casing faces at higher elevations on the pyramid were left "in the rough" (as with G3's granite casing) there would have been nothing for the finishing masons to stand on, that is, if my earlier argument against foothold embankments extending all the way to the ground is correct. I believe that casing stones, except for the lowest courses, were put in place with their outer faces already dressed to the final slope. I also think that for aesthetic reasons the builders would not have left the blocks unfinished during many years of construction. That's just my guess.

Building the arch

In chapter seven I noted that the north and south walls of the Queen's Chamber in G1 do not support the gabled roofing beams. The springs of the arch lie well beyond the walls. In other words, the chamber, at a certain point during its construction, may have looked like that shown in figure 202.

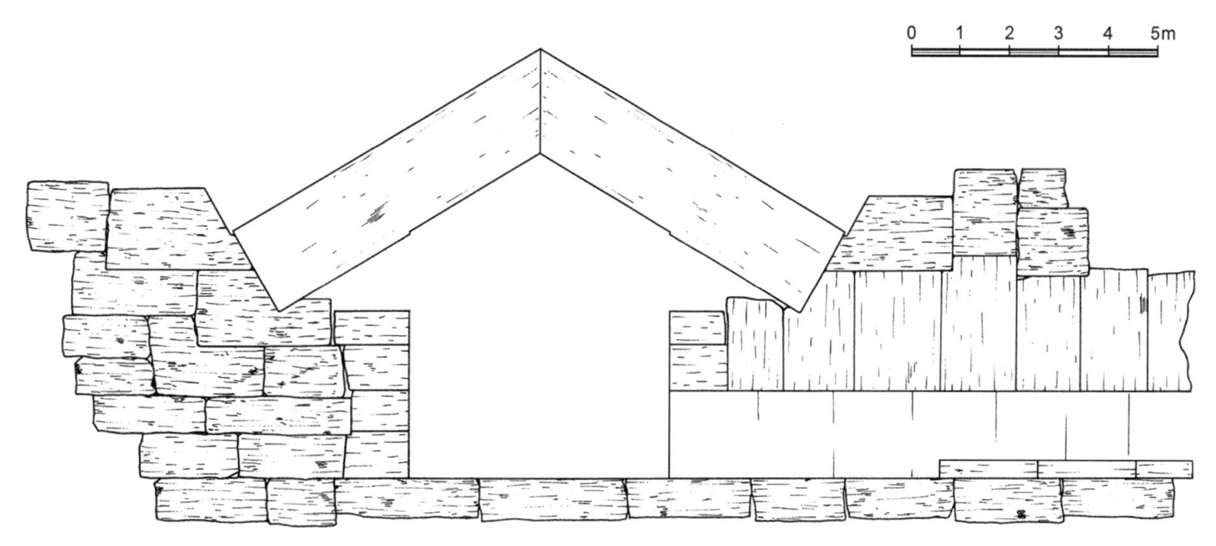

Fig. 202 The arch supports for the roofing beams of the Queen's Chamber.

How were these beams set up? It is well established that masonry arches from Roman times onward were built upon a *falsework*, also called *centering*, usually of wood, upon which the stones (or concrete) were laid. In this way, after the arch is established the falsework can be removed, as the completed arch is self-supporting. It is likely that the arched ceiling of the Queen's Chamber was also built upon a falsework, probably of stone blocks because of the huge weights it had to support. Further, to raise the ends of the beams that would meet in the middle of the ceiling I envision protrusions on the bottoms of the beams so levers could gain purchase in jacking them up.

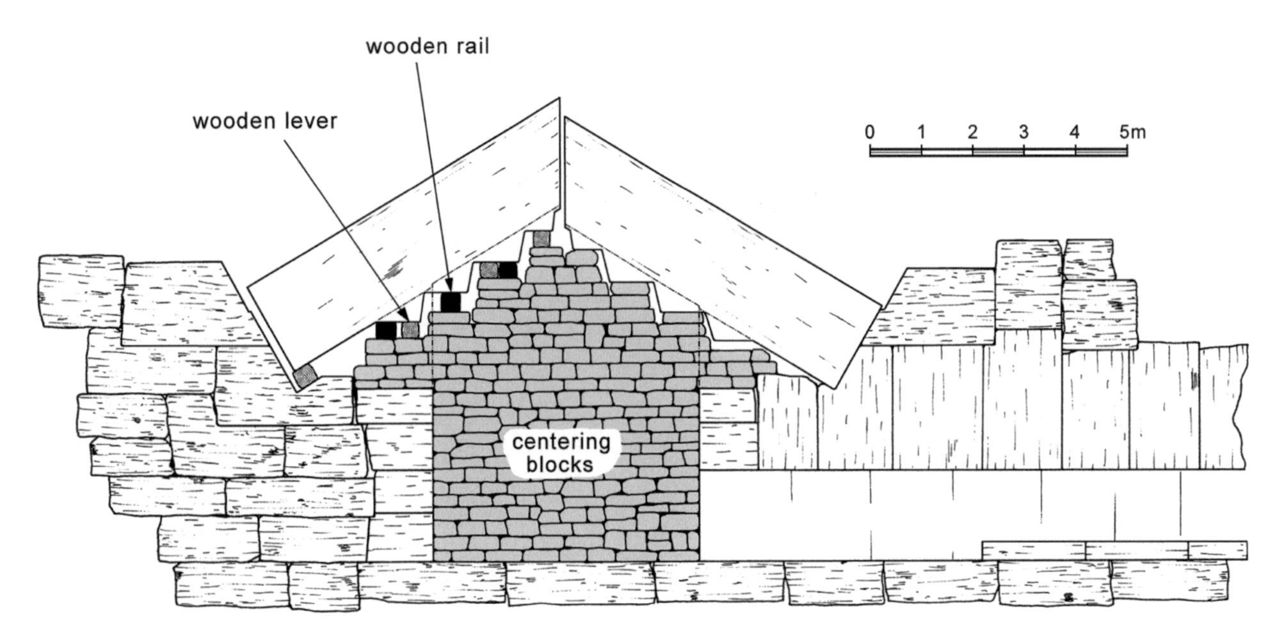

Fig. 203 The centering blocks possibly used in setting up the roof beams of the Queen's Chamber.

This drawing also depicts hypothetical wooden rails that would have allowed the roofing beams, once jacked up, to be slid into position one against the other. After the ceiling was established, the centering blocks could have been safely removed. Then the jacking lumps on the undersides of the beams could have been chiseled away.

Khufu's boat pits

I mentioned Khufu's boat pits in chapter three. Here I will address how the massive limestone beams that formed the ceilings of these pits were put in position, as there is good evidence for the method.

The first scientific report on the eastern pit and its contents was a paper titled *The Cheops Boats*. It was authored by Mohammad Zaki Nour, Zaky Iskander, Mohammad Salah Osman, and Ahmad Youssof Moustafa (Cairo 1960). The physical evidence described here is taken from that paper unless otherwise noted.

The pit was roofed with forty-one massive beams of fine white limestone. The heaviest block measured 4.8 meters long, 1.6 meters high, and 0.85 meters wide (15.7 x 5.25 x 2.8 feet). Its specific gravity was 2.45, yielding a weight of 16 metric tons.

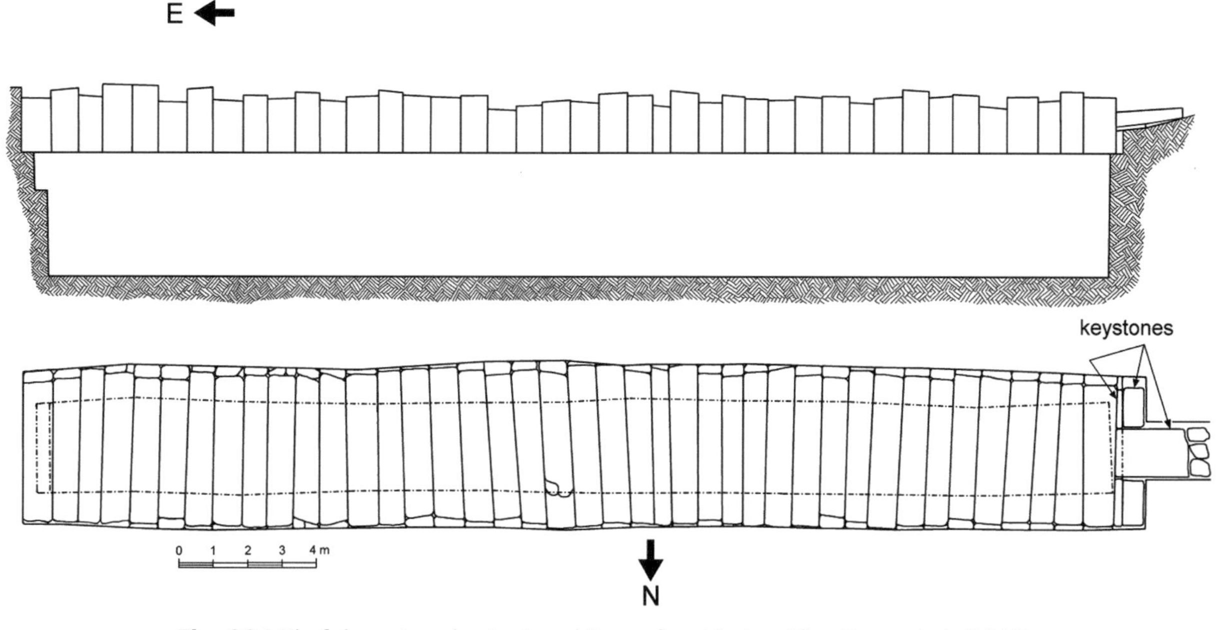

Fig. 204 Khufu's eastern boat pit and its roofing blocks. After Nour, et al, (1960). Redrawn by the author for clarity.

That the beams had been placed from east to west was deduced by engineering director, Dr. Osman, because of the presence of a ramp carved into the rock at the western end of the pit, the ramp containing five "keystones" that secured the westernmost beam.

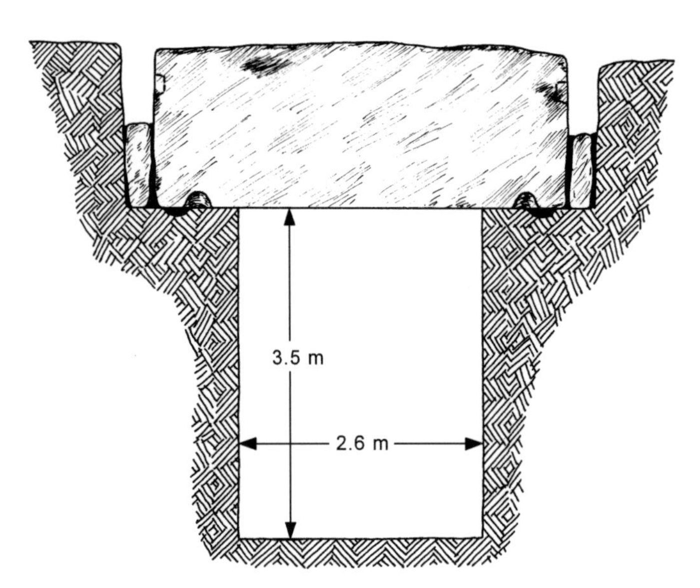

Fig. 205 Cross-section of the eastern boat pit. The boat timbers are not shown.

The roofing blocks rested on shelves about one meter wide cut into the north and south walls of the pit. In cross-section the pit appeared as in figure 205. Between the ends of many of the blocks and the bedrock walls were small slabs of limestone called "shutters" by the researchers. A coarse pinkish mortar filled the larger gaps around the shutters. In the tiny gaps between the roofing blocks themselves was a 99% pure gypsum mortar that had been trickled down between them and even plastered into the seams between adjacent blocks on their undersides.

On removal of the blocks, Osman observed: "A layer of fine sand sometimes mixed with clay was found between the blocks and the horizontal edge of the pit at its support. It is most probable that this layer was applied to slide easily every block in its right place."

Dr. Osman further observed:

On the two horizontal ledges along the northern and southern sides of the pit there were found circular shallow holes at the support of each block. In the top of the eastern vertical edge of the pit to which block No. 41 was fixed two square openings 10 x 15 cms and 10 cms deep were found. Outside the pit towards the east several semi-circular and square holes varying in depth were found. It is most probable that these holes together with those in the pit were used in the operation of levering and adjusting the blocks in their places.

The circular shallow holes in the pit shelves were filled with the same type of coarse pink mortar used to fill around the shutters.

Though the roofing blocks varied in width and height, each had a similar series of incisions, as shown in figure 206. Most blocks had a single rectangular hole at position **A** in both ends of the block. A few of the widest blocks had two holes side-by-side. One narrow block had two holes, one over the other. These holes appear near the tops of the blocks, but a few are closer to the middle. The notches at **B** appear in both ends of each block and were filled with coarse pink mortar. The notches at **C** appear only on the western edge of each block and were also filled with coarse pink mortar.

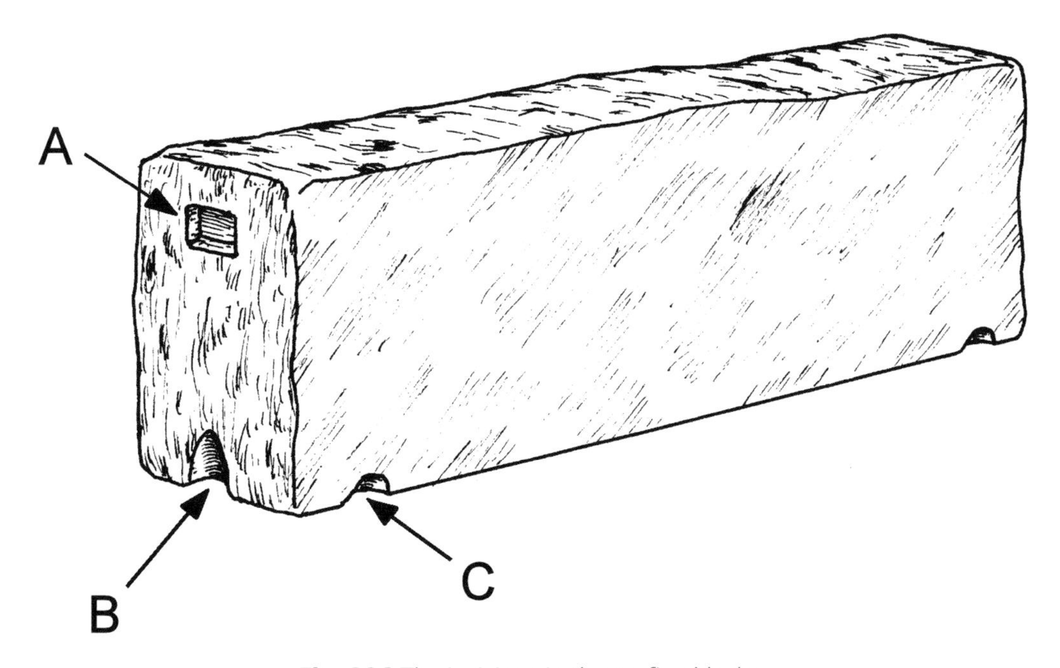

Fig. 206 The incisions in the roofing blocks.

Numerous builders' marks were found on the walls of the pit and on the blocks themselves. Eighteen blocks bore the cartouche of Djedefre, Khufu's immediate successor. Block dimensions in red ochre paint appear on the undersides of sixteen blocks.

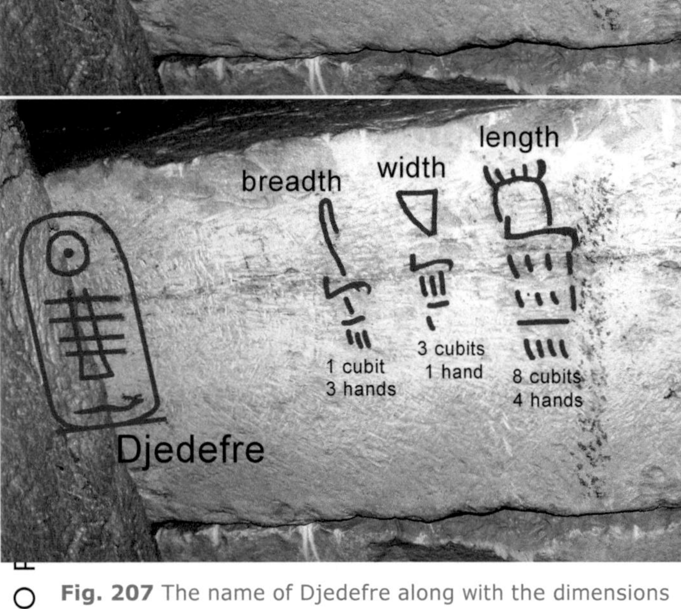

Fig. 207 The name of Djedefre along with the dimensions (width, height, length) of the block in cubits, hands, and fingers. Writing on the lower figure enhanced by author.

As far as I know, no one has hitherto attempted to explain the overall relationship of the pattern of holes and block cuts just described. The regularity of these features block-to-block suggests they are associated with, and are therefore clues to, the placement of these beams.

In figuring out how the blocks were placed it would be helpful if we could determine the order in which the blocks were placed and the location along the pit where each block arrived. The keystones at the western end indicate that the blocks were indeed placed one against the next from the west, as Dr. Osman suggested, but it doesn't necessarily follow that each block entered the pit from the western end. To the contrary, many of the blocks are too wide to fit between the bedrock walls above the shelves at that end of the pit. And, if the blocks entered the western end of the pit in any orientation other than roughly parallel to their final positions (i.e. sideways), maneuvering them would have been greatly complicated. It is more probable that the blocks arrived close to their final location along the pit, then lowered into position.

The natural question would then be: How were the blocks safely lowered to their supporting shelves? To answer this question let us look again at the blocks, then the pit.

Referring again to figure 206, the small crescent-shaped cuts at **C** are characteristic of levering notches seen in many of the casing and backing stones of the first course of blocks in Khufu's pyramid, supporting Dr. Osman's belief that all the roofing stones were pushed from west to east using levers. The notches at **B** seem to be guide slots for ropes, perhaps as large as 6 cm in diameter. The rectangular notches at **A** are, in my guess, lever lifting points.

In the pit itself are found the most important clues to the block placement method, the mortar-filled holes in the block-supporting shelves. Because these holes could have had no function after being filled with mortar and covered by the roofing blocks we must conclude that they served a use before being covered. The function most logically suggested is that the holes were part of the procedure used to lower the blocks. Continuing this train of thought, the holes are well suited for stabilizing the ends of heavily-loaded wooden poles, keeping them from slipping sideways.

With these premises I am able to suggest the method used to lower the blocks:

1. A sled-borne block, lashed to its transport sled, is dragged to the position shown in figure 208. The lashing is removed. The holes (circular depressions) in the shelves used for the previous block are filled with mortar. The holes in the shelves needed for this new block are cut.

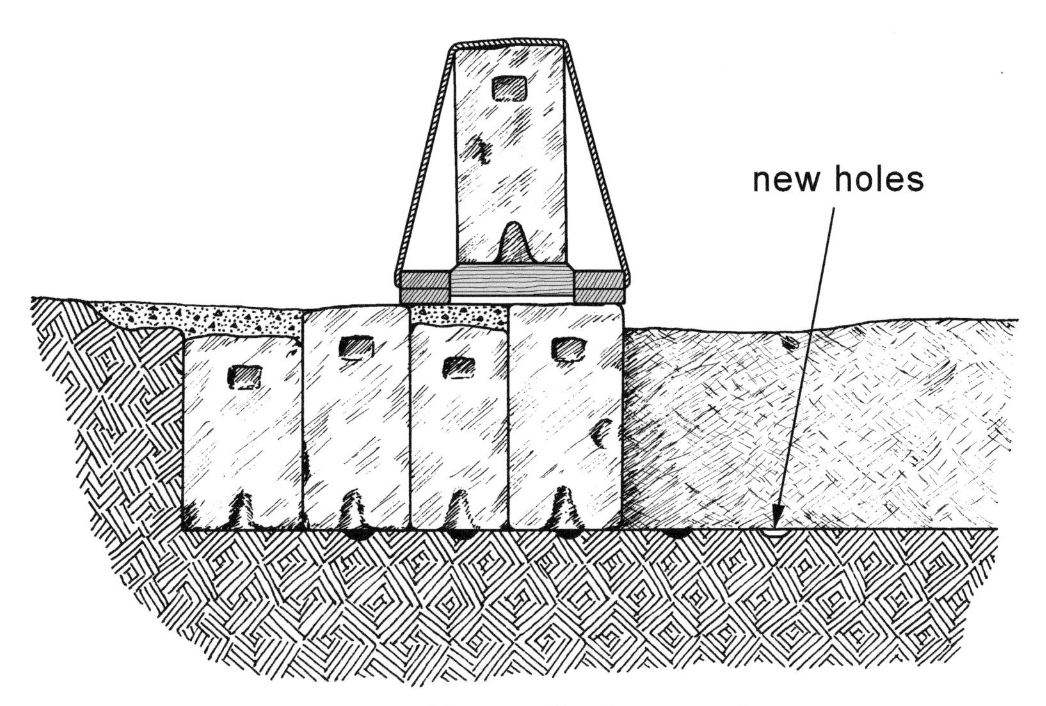

Fig. 208 The sled-borne roof block arrives at the pit.

2. An A-shaped wooden frame is positioned with its feet finding the holes in the shelves. Lifting ropes are attached to the block, and tension is applied to the ropes.

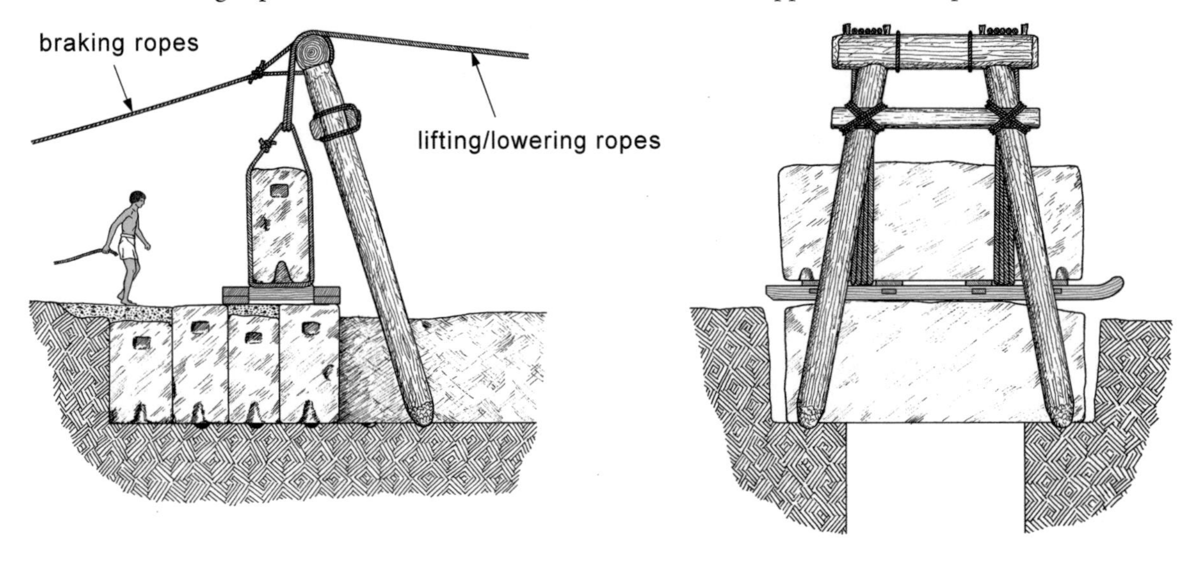

Fig. 209 The A-frame is positioned with its feet secured in the shelves.

3. Increasing tension on the ropes by workers to the west makes the A-frame rotate, lifting the block from its sled and positioning it directly over its final location, as shown in figure 210.

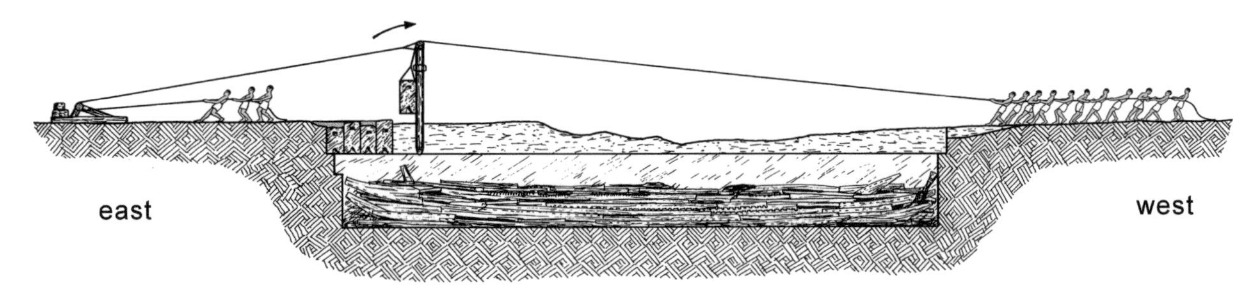

Fig. 210 Workmen to the west pull the A-frame to vertical.

During this operation the lifting ropes will not slide relative to the horizontal wooden beam over which they pass because friction between the ropes and beam, and other forces, are in balance. Braking ropes attached to the A-frame, and running around anchoring means to the east, prevented the A-frame from rotating past vertical, a situation that would have been disastrous.

Calculations show that the mechanical advantage of this rotating A-frame is such that the maximum force required of the workmen to the west is about half the force required to lift the block off its sled.

4. The workers to the west relax tension on their ropes and allow the block to be lowered to within about 15 cm of the shelves. As long as tension on the braking ropes to the east is maintained, the A-frame will remain vertical (not rotate) during this phase. Also, because rope-to-beam friction is advantageous to the workmen lowering the block, the tension in the ropes on their side of the beam is only two thirds of the tension in the ropes from the beam to the block. I have verified the workability of this procedure both mathematically and with a small model.

The notches at **B** (fig. 206) in the ends of the blocks may have held a rope employed by several workmen at each end of the block to make sure the block was lowered evenly (fig. 211).

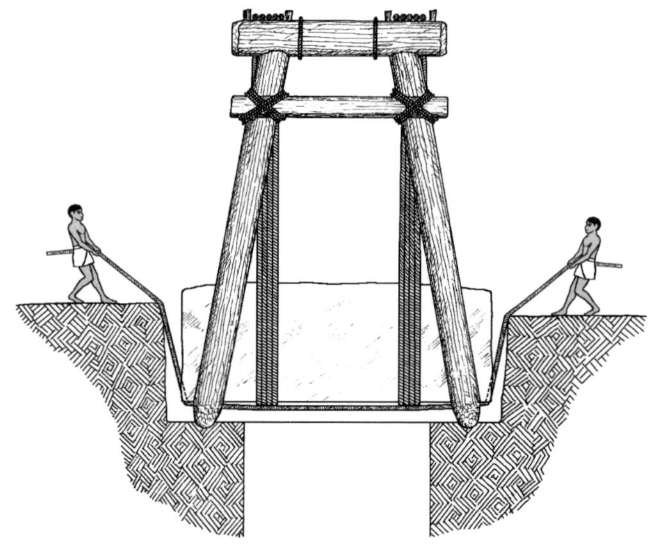

Fig. 211 Workmen ensure the block is lowered evenly.

5. The safety rope is withdrawn and the block is lowered the last few centimeters to the shelves.
6. The lifting/lowering ropes are removed.

7. The A-frame is moved away a couple of meters so that workmen can insert levers into the slots at **C** (fig. 206) and push the block flush to its neighbor, the pushing force being substantially reduced by the layer of sand and clay on the shelves.
8. The notches at **B** and **C** (fig. 206) are filled with mortar. Shutters are placed at the ends of the block and mortar is filled around them, locking the block in place. Pure gypsum mortar is poured into the cracks between the roofing block just placed and the previous block.
9. Steps one through eight are repeated for the remaining blocks.

A hypothesis is strengthened if its predictions are confirmed. One of the predictions I made before seeing the photograph (figure 212) below is that there would be no circular depressions in the rock shelves under the first block put into the pit, the easternmost block. To lower the first block the A-frame legs had to be where the second block would be located. In 1987 one of the roofing blocks sat in the far eastern position, so I was unable to observe whether there were holes in the shelves underneath. However, when I subsequently found the photograph just mentioned, it appears that there were no holes under the first block. Note: I suspect the hole locations in figure 212 were retouched by Nour, et al, to make them more distinct; the holes are not nearly so obvious, especially since most are mortar filled.

Fig. 212 The holes and notches at the east end of the pit. Photograph from Nour, et al, (1960). Arrows and note added by the author.

The procedure I have described may be inaccurate in detail (e.g. rope rigging and exact shape of the A-frame), but I believe it is a reasonable explanation of the features we observe. Would the A-frame have had other applications? It may have been used for loading and unloading cargo boats, as we do not have a good theory on how this was done.

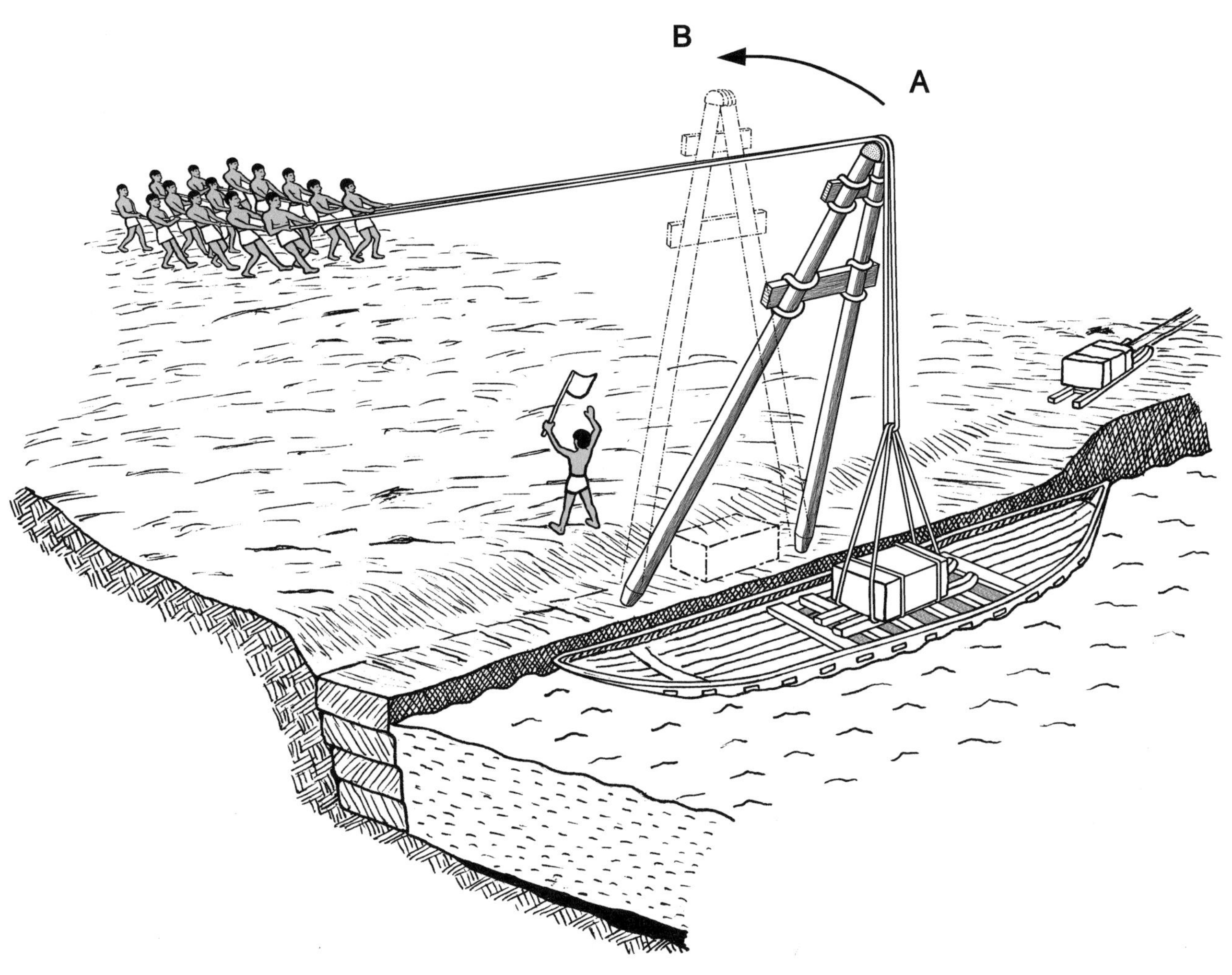

Fig. 213 Unloading a block from a cargo boat.

I doubt that the A-frame was employed for raising pyramid blocks because we do not see pairs of holes anywhere on the pyramid that would indicate its use.

Sealing pyramid passageways

Most of the inclined passages in the stone pyramids were built at slopes that yield angles between 24 and 30 degrees. For example, Petrie found the mean angle of the Descending Passage of the Great Pyramid to be 26.523 degrees, a slope of one rise to two run. Because the northern entrance corridors point generally toward the celestial North Pole they have been called "polar passages," implying religious or scientific purpose for their orientation. If this was true, how do we explain virtually the same passage slopes in the western corridor of the Bent Pyramid and Ascending Passage in Khufu's pyramid? Perhaps the passages were inclined for another reason.

To investigate this problem I positioned a limestone slab on my friction measuring device (described in chapter three) at the corridor slopes found in the pyramids. I placed cubes of granite and limestone on the slab. The results were enlightening. When the surfaces in contact are dry and clean the cubes cannot hold their positions at angles of 25 degrees and 26 degrees, or greater, for granite and limestone respectively. They accelerate down the slab.

If I rub either cube against the slab I produce a light coating of stone powder between them. Each cube will now stick to the slab at the same angle where previously it lost its grip. However, a slight push will move the cube down the slope. The cube stops if I stop pushing. Thus it appears that under the right conditions one person could push a stone block weighing several tons down an inclined corridor such as noted above. My proposition: The passageways were sloped for the practical purpose of controlling the sliding of plug-stones in sealing the pyramids.

The trial passages

One last example of engineering I find remarkable is the feature carved in the bedrock on the east side of G1 called the "trial passages." It is a full-scale near-replication of the junction of the ascending and descending passages in Khufu's pyramid, including the lower end of the Grand Gallery and the notches in its floor where the wooden bridge would cross. These passages, if created before the body of G1 was commenced (we don't know if this was so), would be additional evidence against the change-of-plan theory that I argued against in chapter seven.

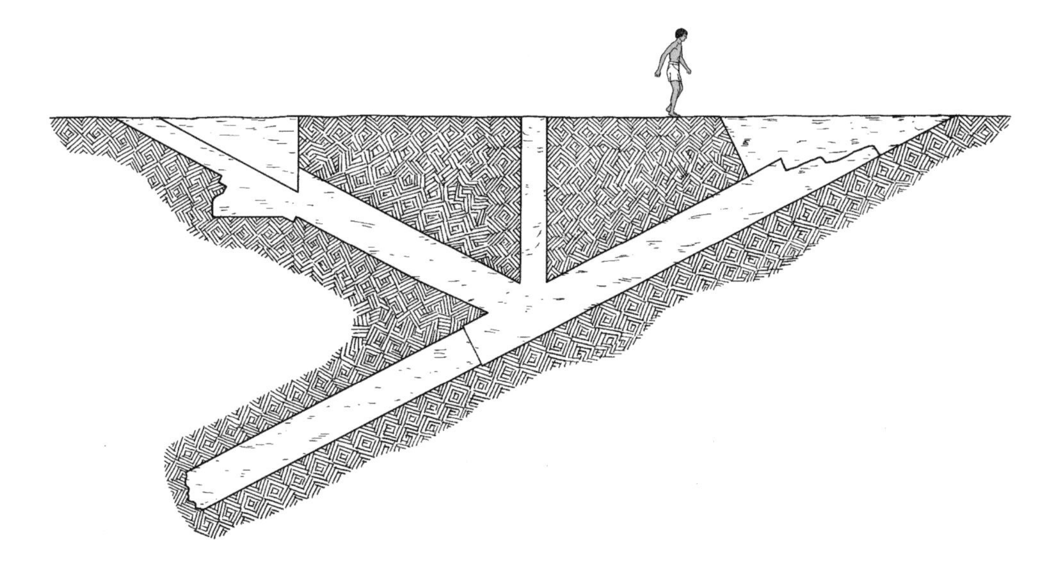

Fig. 214 The trial passages. Redrawn by author from Petrie (1885, pl. ii).

Mark Lehner believes this feature helped control the building of the main pyramid. He cites the near-coincident east-to-west alignment of the vertical tunnels in the well shaft of G1 and that in the trial passages. He further proposes that the trial passages were to be incorporated into a small version of Khufu's main pyramid, a project that was abandoned in its early stages.

Lehner's hypothesis is intriguing, but unlikely. The vertical tunnel of the trial passages is not aligned with either of the two vertical portions of the well shaft in G1. Nor can its hypothetical counterpart be seen at the junction of the ascending and descending passages. Petrie said of the vertical shaft in the trial passages (1885, 16):

The vertical shaft here is only analogous in size, and not in position, to the well in the Pyramid gallery; and it is the only feature which is not an exact copy of the Great Pyramid passages, *as far as we know them* [emphasis by Petrie]. The resemblance in all other respects is striking, even around the beginning of the Queen's Chamber passage, and at the contraction to hold the plug-blocks in the ascending passage of the Pyramid.

I believe the trial passages are too distant and too misaligned from their counterparts in the pyramid to have served as control means. Further, there is not sufficient evidence that a pyramid was planned at this spot — no foundation work can be seen. I think the builders made these passages to study the sealing system for the pyramid, an indication of meticulous planning.

10

RAISING THE STONES

Much so far has been information perhaps useful in figuring out how the stones were raised. In this chapter I will examine stone raising theories. But first, consider a few more observations and bits of evidence that might help.

Local quarry sites

As noted earlier, the Klemms identified specific quarry locations for most of the Old Kingdom pyramids. But I will focus on Giza for the same reason I gave in chapter one: more excavation has been done here, so more is known.

The G3 quarry is the easiest identified of the Giza pyramids. It lies only a hundred meters southeast of the pyramid. Reisner said it could have supplied all of the core blocks for G3. The Klemms, supporting Reisner, say (2010, 98) "The geochemical correlation diagrams of the Menkaure pyramid's core material unequivocally assign the stones to the quarry area situated south-east of the pyramid."

I can argue similarly for G2. The "Great Quarry" southeast of G2 is positioned relative to G2 as the G3 quarry is to G3 (see fig. 6). Thus it is *a posteriori* implied that this quarry served G2 in addition to G1.

Fig. 215 The Great Quarry southeast of G2.

The north and west edges of this pit appear as a rock-cut escarpment. The southeast limit is not obvious, but must extend well beyond the tomb of Khent-Kawes; bedrock rises to mid-level of her stepped mastaba. Lehner gives dimensions of this "bowl" as 400 meters (1312 feet) north-to-south by 300 meters (984 feet) east-to-west. Its lowest part is about 25 meters (82 feet) below its highest edge. Lehner believes this quarry supplied most of the stone for G1. Hawass agrees. The Klemms found that the eastern part of the Great Quarry served G2.

The bottom of the Great Quarry has never been uncovered. The reason? Hawass and Lehner say ("Builders of the Pyramids," 1997, 33) "This quarry, as well as areas later mined by Khafre and Menkaure, is filled with millions of cubic yards of limestone chips, gypsum, sand, and *tafla* [desert clay]. This debris, probably remains of construction ramps, was dumped into the quarry after the pyramid [Khufu's] was finished."

There is also a shallow quarry between the row of Dynasty 5 mastabas south of G1 and north of the G2 causeway. It probably supplied G1 in its early building phase.

Why are quarry locations important? Because the quarries are found at singular locations with respect to each pyramid served. That is, stones were not brought from locations all around each pyramid; they came from one general direction. This may be a clue in the stone-raising method.

Block size vs height raised on the pyramids

Another consideration in block-raising is how blocks vary in size with respect to their heights in the pyramid. Block size tends to diminish with height – larger blocks down low and smaller blocks up high. While generally true, it is not always the case. For example, looking at figure 144, the upper portion of G2, notice *casing* courses that are substantially thicker than courses above and below. Also, just below the casing are *core* blocks not much smaller, if any, than those lower. Another example is the unusually large granite casing stone on the west face of G3 (fig. 196) about a third of the way up the pyramid. It appears the block-raising method was not overly influenced by block size.

Ramp evidence

One might suppose that the best places to look for ramp evidence would be at unfinished pyramids. Four Old Kingdom pyramids were abandoned before completion. In order of construction they are:

1 – Sekhemkhet, Dynasty 3, Saqqara
2 – Sinki, Dynasty 3, Abydos
3 – Layer Pyramid, Dynasty 3, Zawiyet el-Aryan
4 – Djedefre, Dynasty 4, Abu Rawash

A fifth pyramid, Menkaure's at Giza, has unfinished granite casing on its lower portion, but was near enough to completion that its ramps, if used, were likely removed.

At the Layer Pyramid and that of Djedefre no remnants of construction ramps were found. Maragioglio & Rinaldi say of the Layer Pyramid: "And we think it important to mention the fact that, although the pyramid was not finished, it had reached a certain height: and yet no traces exist of a possible frontal construction ramp of the type suggested by Lauer . . ." The same authors say of Djedefre's pyramid: "If a working ramp had been used and erected for its construction, traces of it at least would certainly have remained, since the pyramid had reached a certain height and the clearing up of the complex at the King's death was hurried and summary. The topography of the ground and the other buildings surrounding the monument leave few places where such a ramp could have been made, but no remains that can reasonably be attributed to this auxiliary construction are visible in the area."

As for the small unfinished Dynasty 3 pyramid of Sinki (at Abydos), Dieter Arnold said (1991, 81):

. . . four ramps lead from all sides against the inclined faces of the pyramid. Each ramp is 12 meters long and has an angle of about 12 to 15 degrees. These ramps, if completed, would have reached a height of only 6 meters. To raise the material to the top of the pyramid, which was planned to be 12 meters high, considerable additions would have been necessary. Since the ramps cover the lower parts of the third layer, they would have been partially dismantled and rebuilt, similar to the ramp [Sekhemkhet] at Saqqara. This defect seems to indicate the still-limited experience in the early period of pyramid building.

I don't agree that these ramps represent an experience "defect" or poor planning. I think it is likely that blocks for higher levels were *carried* up these ramps to the pyramid faces and then hoisted to higher levels.

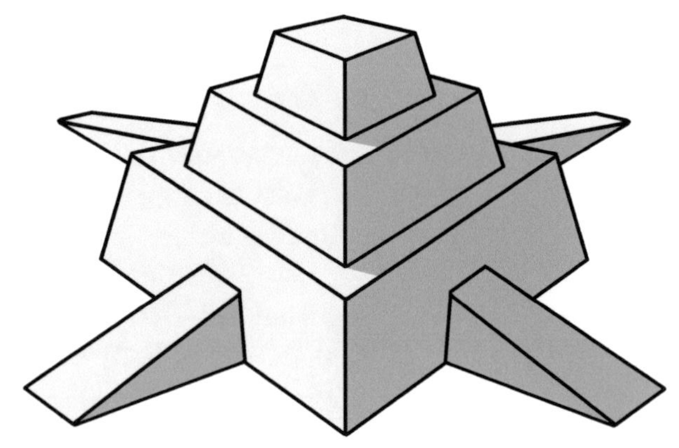

Fig. 216 The Sinki pyramid (if completed)

Sekhemkhet's pyramid is puzzling. Goneim first thought that the embankments covering the pyramid, and especially the large supply ramp on its western face, were used for its construction. Later he opined that the embankments were used to bury the site. Maragioglio & Rinaldi reinforced Goneim by reasoning that a supply ramp would not have risen higher than the pyramid side against which it laid.

Now we are in a quandary. We have four pyramids abandoned at about the same stage of construction. Two display not a smidgen of ramp evidence, one has a ramp serving each of its four sides, and one is perhaps covered with ramp material. If ramps were employed at the Layer Pyramid and Djedefre's, and the work raising these pyramids stopped abruptly, why did the builders take time to remove the ramps? Were the embankments at Sekhemkhet's pyramid construction aids, a cover-up of the site, or both? It is impossible to know. But one thing we can conclude from Sekhemkhet's pyramid: the Egyptians could move a lot of dirt!

Menkaure's pyramid offers tantalizing clues, if only we could interpret them. First, because it was not quite completed, we are allowed a glimpse of unfinished casing. One could argue that its granite casing is atypical, but still we must explain how casing in this aspect (unsmoothed, no drag marks on outer surfaces) was erected. Second, the pyramid has a nucleus structure of large steps, so the stone-raising method has to account for that. Third, its singular quarry to the southeast means that a big ramp, or multiple ramps, remain feasible – if stones approached the pyramid from all directions a single large ramp would seem less likely.

As for the approach road at Meidum (fig. 76), Wainwright suggested it was used to bring building materials to the site, particularly for construction of the stepped pyramids E1 and E2. This road is straight, and points just south of the center of the east face of the pyramid. The road is built on a gradient that would take it to the base of the pyramid. Because of its proximity to the pyramid and its low gradient it seems improbable that this road was part of a stone-raising ramp.

In chapter six I mentioned the drag ramps southwest of the Red Pyramid. These appear similar to the Meidum approach road. Core stones were likely brought via these ramps, but there is no evidence that the ramps were extended to higher elevations.

Dieter Arnold describes what he believes were traces of construction ramps at the Middle Kingdom pyramid of Senwosret I at Lisht:

They consisted mainly of bricks and Nile mud, retained by side walls of bricks. In one case (at the ramp in the center of the west side), the bricks were laid at an inclination of 12 degrees, which must have been the slope of the ramp. Remains of a second ramp approach the pyramid from the south, near the southwest corner. Since both ramps seem to have had a length of only about 50 meters, they could—with moderate slope—have reached only the lower parts of the 60-meter-high pyramid.

Examples of ramps with even steeper slopes have been reported. Again from Arnold:

Under the pavement of the court of the sun temple of Niuserra [Dyn 5] at Abu Ghurab, east of the obelisk, Borchardt found five 2.5-5-meter-thick brick ramps, which fanned out from the obelisk. Their sides were vertical, and the bricks were set horizontally. But from a red line on the outer face of the obelisk base, Borchardt concluded that the ramps had an inclination of 14 degrees. They interrupt the foundations of the eastern court wall and were certainly construction ramps for the obelisk.

Dr. Arnold also tells us of small stone ramps as steep as 18 degrees that were used to haul up the heavy lids of sarcophagi in the burial chambers of Ti and Mereruka at Saqqara. Ramp inclinations at least that steep were observed by Reisner and Herman Junker at Giza.

Against unfinished pylon #1 of the Great Temple at Karnak (New Kingdom) are remains of a construction embankment having rather steep faces. The embankment is reinforced by a network of rubble walls similar to those that surround Sekhemkhet's pyramid. It is not obvious that the working platform was ever supplied by an inclined ramp. Could these embankments have supported some kind of hoisting device? We don't know.

Machine evidence

No machines or obvious fragments of machines have been found at any pyramid. That is not surprising. We should not expect to find machines or their components. Most likely of wood, machines would have eventually succumbed to the ravages of time, spent pieces ultimately serving as firewood.

What about marks on the pyramids that machine use may have left? Some of the block cuts I mentioned in the last chapter could be evidence of some sort of device. But they appear so infrequently that they are unlikely to have been part of a general stone-raising procedure.

We also have no evidence of a stone-raising machine in drawings, paintings, or models.

Other than the A-frame I hypothesized in the last chapter, we have no evidence of a machine more complex than the lever. As Clarke & Engelbach said (1930, 95), "The more, however, his constructional methods are studied, the more one is convinced that if any detail in a piece of work has to be explained by an apparatus of any complication, then that explanation is certainly wrong."

Rope evidence

Most stone–raising theories include use of ropes.

The Egyptians made ropes from several vegetable fibers. According to Alfred Lucas (1962, 135), the most common material was date palm fiber. The only ropes dated with certainty to the Fourth Dynasty, however, those found with Khufu's royal ship, were made from halfa grass. Several examples of ancient rope may be seen in the Cairo Museum.

Lucas also reported:

In May 1942 seven very thick ropes were found buried in debris in one of the Tura caves, which are old stone quarries. These ropes were of papyrus and consisted of three strands, each of which had about forty yarns and each yarn about seven fibres. The circumference was about eight inches and the diameter about two and a half inches. They are not modern, but the date is uncertain.

Incidentally, a good quality hemp rope of this size has a minimum breaking strength of 34,000 pounds - seventeen tons!

Evidence for ropes in building construction is plentiful. Huge portcullis slabs that sealed Dynasty 1 and 2 pit tombs have rope holes, and include grooves that allowed the ropes to be withdrawn when the slabs were finally positioned.

Fig. 217 Rope-lowered portcullis blocking burial chamber of Tomb 3035 at Saqqara. From Emery (1961, plate 16).

The four-ton crypt-sealing granite cork in Djoser's pyramid and the roofing stones for Khufu's boat pits were obviously lowered with the help of ropes. Portcullis gates in the pyramids of Khufu and Menkaure were apparently lowered by ropes wrapped around logs for which slots in the chamber walls appear.

Ropes not only rigged Egyptian ships but also held many parts of the vessels together, especially the hull planks. On large cargo vessels like Hatshepsut's obelisk barge (fig. 56) the hogging brace of twisted ropes may have borne tensile loads upward of a hundred tons.

Hassan's stone bearings

There is rope evidence that could be pertinent to stone-raising. Two identical stone objects were discovered by Selim Hassan while excavating at Giza in 1932-35. Hassan called these objects "pulleys." He found one in a mud-brick house near the tomb of Queen Khent-Kawes, the other while cleaning an area east of G2 (Hassan 1960, X, 49). The stone artifacts are wrought from red Aswan granite. Each contains three shallow, parallel, polished grooves that suggest tracks for ropes of about 28 millimeters (one-and-one-eighth inch) diameter.

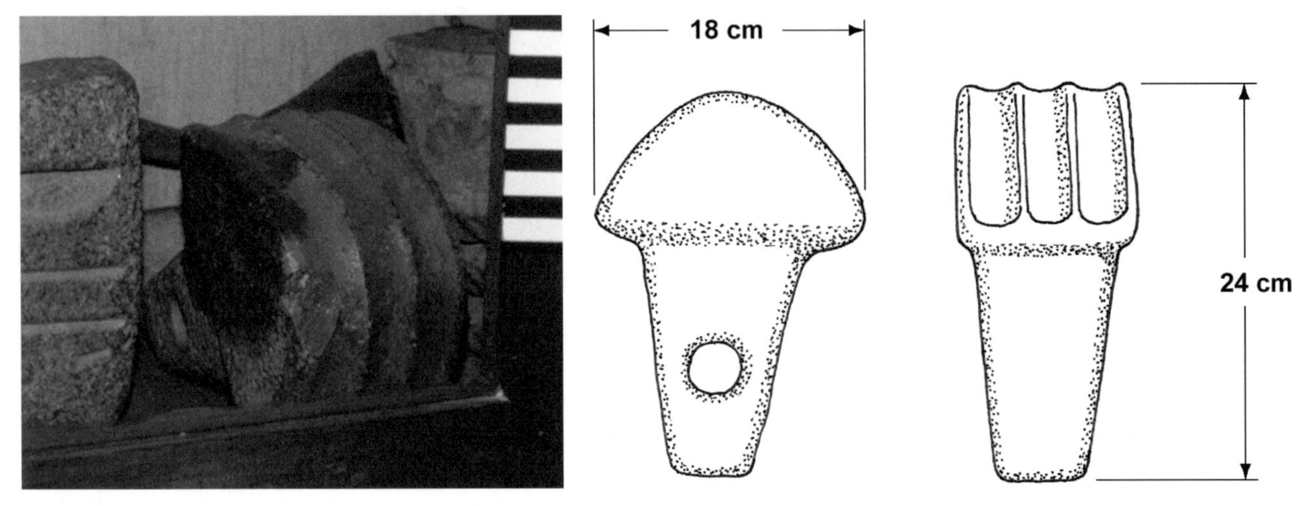

Fig. 218 Hassan's stone pulley (Cairo Museum).

Fig. 219 Hassan's stone pulley.

Opposite the grooved face is a tapered stem pierced by a single hole. The stem appears to be a tenon which would be retained in a mortised slot by a peg.

This device deserves attention, for it could be a key to raising the stones.

In situations where a heavy load is being pulled by a rope it is sometimes helpful to arrange the rope such that it is not being pulled in the same direction that the load is moving. This change of rope direction, at least since Roman times, has been accomplished by running the rope around a small wheel-and-axle device called a *pulley*. The familiar use of the pulley is in *lifting* loads vertically, but in a more general case it greatly reduces friction, or energy lost, if the rope must go around the corner of a fixed surface. If a pulley is not available, or is unknown, a *plain bearing* having a smooth, polished surface is the next best thing. A plain bearing is what Hassan's pulley is.

Hassan's granite bearing would be particularly useful for hoisting and maneuvering blocks in tight places. The reasons are:

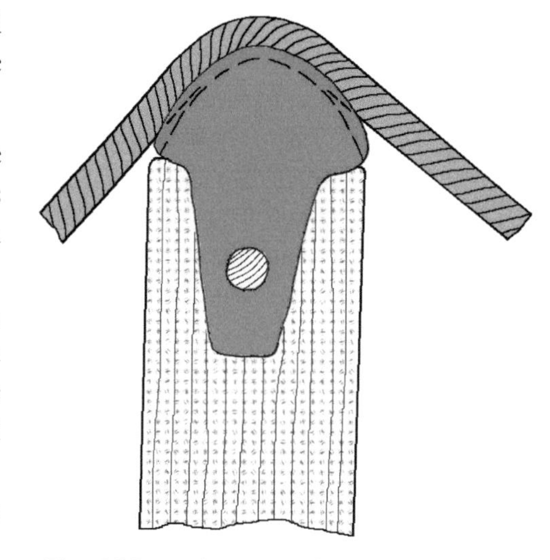
Fig. 220 Employment of Hassan's bearing.

1. However used, the bearing was designed to reduce friction produced by ropes going around a corner. It was not used for *lowering* weights because for that operation a friction-reducing bearing is a hindrance.
2. The size of rope used by this bearing is proper for moving, including lifting, a pyramid block of average weight with only two ropes. An extra rope would have increased the margin of safety. Heavier blocks would have required more ropes and thus more than one bearing.
3. Because the stone bearings were found at Giza they were probably contemporary with building of the pyramids.
4. The strength and wear resistance of Aswan granite would have been well suited for this application.
5. That the two bearings were identical implies a *standard* design, suggesting that more existed.

Anecdotal evidence

Greek historian Herodotus (484-425 BCE) provides the earliest account of pyramid building in Book II of *The Histories* (quoted from *The History of Herodotus*, 1932):

… Cheops succeeded to the throne and plunged into all manner of wickedness. He closed the temples, and forbade the Egyptians to offer sacrifice, compelling them instead to labour, one and all, in his service. Some were required to drag blocks of stone down to the Nile from the quarries in the Arabian range of hills; others received the blocks after they had been conveyed in boats across the river, and drew them to the range of hills called the Libyan. A hundred thousand men laboured constantly, and were relieved every three months by a fresh lot. It took ten years' oppression of the people to make the causeway for the conveyance of the stones, a work not much inferior, in my judgment, to the pyramid itself. This causeway is five furlongs in length, ten fathoms wide, and in height, at the highest part, eight fathoms. It is built of polished stone, and is covered with carvings of animals. To make it took ten years, as I said - or rather to make the causeway, the works on the mound where the pyramid stands, and the underground chambers, which Cheops intended as vaults for his own use: these last were built on a sort of island, surrounded by water introduced from the Nile by a canal. The pyramid itself was twenty years in building. It is a square, eight hundred feet each way, and the height the same, built entirely of polished stone, fitted together with the utmost care. The stones of which it is composed are none of them less than thirty feet in length.

The pyramid was built in steps, battlement-wise, as it is called, or, according to others, altar-wise. After laying the stones for the base, they raised the remaining stones to their places by means of machines formed of short wooden planks. The first machine raised them from the ground to the top of the first step. On this there was another machine, which received the stone upon its arrival, and conveyed it to the second step, whence a third machine advanced it still higher. Either they had as many machines as there were steps in the pyramid, or possibly they had but a single machine, which, being easily moved, was transferred from tier to tier as the stone rose - both accounts are given, and therefore I mention both. The upper portion of the pyramid was finished first, then the middle, and finally the part which was lowest and nearest to the ground.

This is an instructive chronicle. But is it factual? How did Herodotus come by this info? Earlier in the same section on Egypt he says: "Thus far I have spoken of Egypt from my own observations, relating what I myself saw, the ideas that I formed, and the results of my own researches. What follows rests on the accounts given me by the Egyptians, which I shall now repeat . . ." This makes it obvious that Herodotus was reporting the building of the pyramid as *hearsay* by people living more than two thousand years after the events occurred.

Much of what Herodotus was told is obviously wrong (the pyramid and block dimensions, the underground island, no mention of local quarries). Still, these inaccuracies may not negate the parts that could be true. For example, there could be something to the "machines formed of short wooden planks." Unfortunately, our imaginations must work overtime to envision these devices, given no particulars of design.

There are many other pyramid references by Greek, Roman, and Arab writers, but all appear as speculative as more modern guesses.

That's about it for stone-raising evidence, physical and anecdotal, I think meaningful. We are now ready to examine stone-raising hypotheses that have been proposed over the years. As I look at each I will, after describing it, offer my thoughts on its likelihood of being correct. For variety I use the words theory, idea, suggestion, proposal, and hypothesis interchangeably. Each of these words sound more studious than "guess," but guesses are what they are.

Criteria I think any stone-raising hypothesis must satisfy are these:

1. The proposal must be scientifically valid. It cannot violate physical laws as we understand them. It must not rely on supernatural, extraterrestrial, or paranormal forces.
2. The proposal must be complete enough to explain the most difficult problems. Examples are raising the 70-ton roofing stones of the King's Chamber in G1 and the pyramidion (capstone) on any pyramid.
3. The idea must be in harmony with historical time, materials, and labor available. The time constraint must allow the six largest pyramids (Meidum to Menkaure) to be built within a hundred years.
4. The proposal must be consistent with visible evidence. For example, it must accommodate the dead-straight hip lines on the Bent pyramid and at the top of G2.
5. The proposal should not require technology unlikely possessed by the builders, or complex technology for which there is no evidence.
6. The proposal must be in harmony with known quarry locations.
7. The proposal must account for nucleus structures and all other internal features of the pyramids.
8. The stone-raising method must be relatively safe for the workers. It cannot be so dangerous that many workers would be injured every day.
9. There must be no *negative evidence* regarding the proposal, evidence we would expect to see if the theory is correct, but in fact we do not see.
10. The proposal must be testable, at least in principle. "In principle" means you may not have the resources to test your idea, but can imagine it could be tested if you did.
11. The idea, like any scientific theory, must be falsifiable, capable of being proved wrong. An example of a non-falsifiable theory is that alien beings raised the stones.

Let's see how stone raising theories comport with these criteria. I'll cover machine theories, hoisting, ramps, and a few oddballs. Finally I will present my best idea (spoiler: I would not bet my life on it).

MACHINES

Machine theories envision some kind of wooden crane, stiff-leg derrick, rocking beam (an elevated lever), or levers in combination with cribbing. *Cribbing* is a stack of wooden planks that supports a load such that the load can be raised a little at a time.

The rocking beam

Lewis Croon envisioned a rocking beam that could be moved along the pyramid courses. Ropes attached to its outer arm allow workers greater access, and thus leverage, than would be possible otherwise.

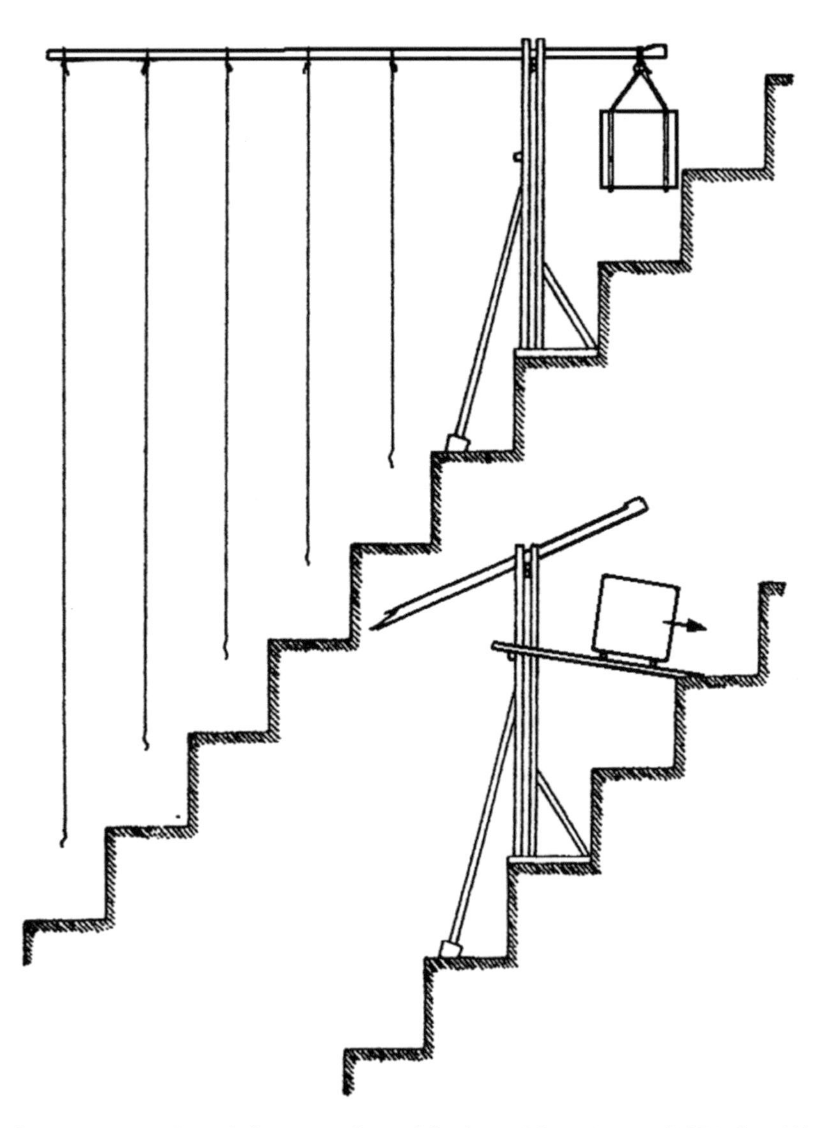

Fig. 221 Lewis Croon's "Device of wood." Adapted from Lauer (1974, fig. 60).

At broad brush, the rocking beam seems not only reasonable, but has precedent. Egyptians have used a similar device, called a *shadoof*, to raise water since ancient times.

Fig. 222 The shadoof.

Problems occur, though, when we increase the scale of this appliance to handle six-ton monoliths. The beam machines would have to be much larger and heavier than proponents envision. To lift six-ton blocks like those on the ninety-eighth course of Khufu's pyramid (about halfway up) the beam machines would have to be at least as large as shown in figure 223. They would weigh about four tons, and could lift blocks only one course at a time. The machines needed to raise the 10-ton corner-casing blocks of the 35th course would have been larger still.

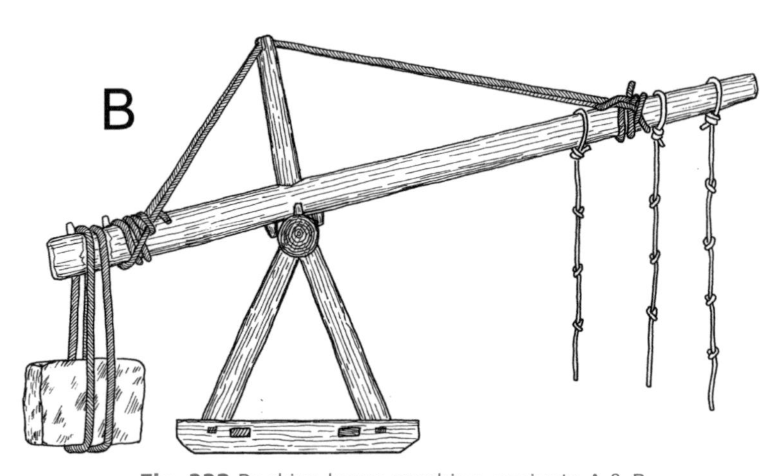

Fig. 223 Rocking beam machine, variants A & B.

Unless the lever arm of our beam is substantially longer than shown in my drawing, more than fifty men of average weight are needed to counterbalance a six-ton block. Where do we place the machines and their operators? The pyramid faces are steep. Even if the casing stones were left as perfect stair-steps, to be dressed smooth later, the average tread width is less than a half-meter. Substantial platforms of stone would have to be erected on the pyramid face for each machine. Anchoring these platforms to the pyramid would not be easy. The only pyramid having unsmoothed casing is that of Menkaure, and on its faces I assure you no platforms would stick.

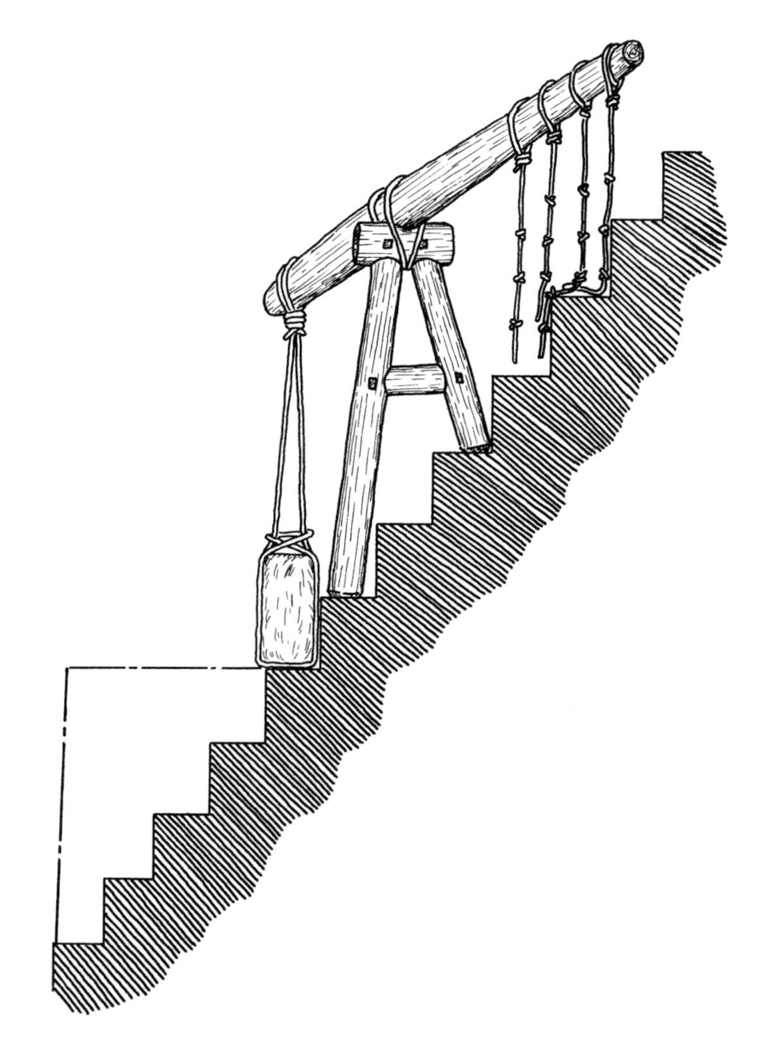

Fig. 224 Rocking beam machine on steep pyramid face.

The stiff-leg derrick

Roman builders employed A-frame and stiff-leg derricks to raise building stones. Architect Marcus Vitruvius Pollio (late 1st century BCE to early first century CE), in Chapter two of Book 10 of *De Architectura*, gives a detailed description of an A-frame derrick called a *trispast*. The trispast employed three pulley wheels, or sheaves. The *pentaspast* used five sheaves.

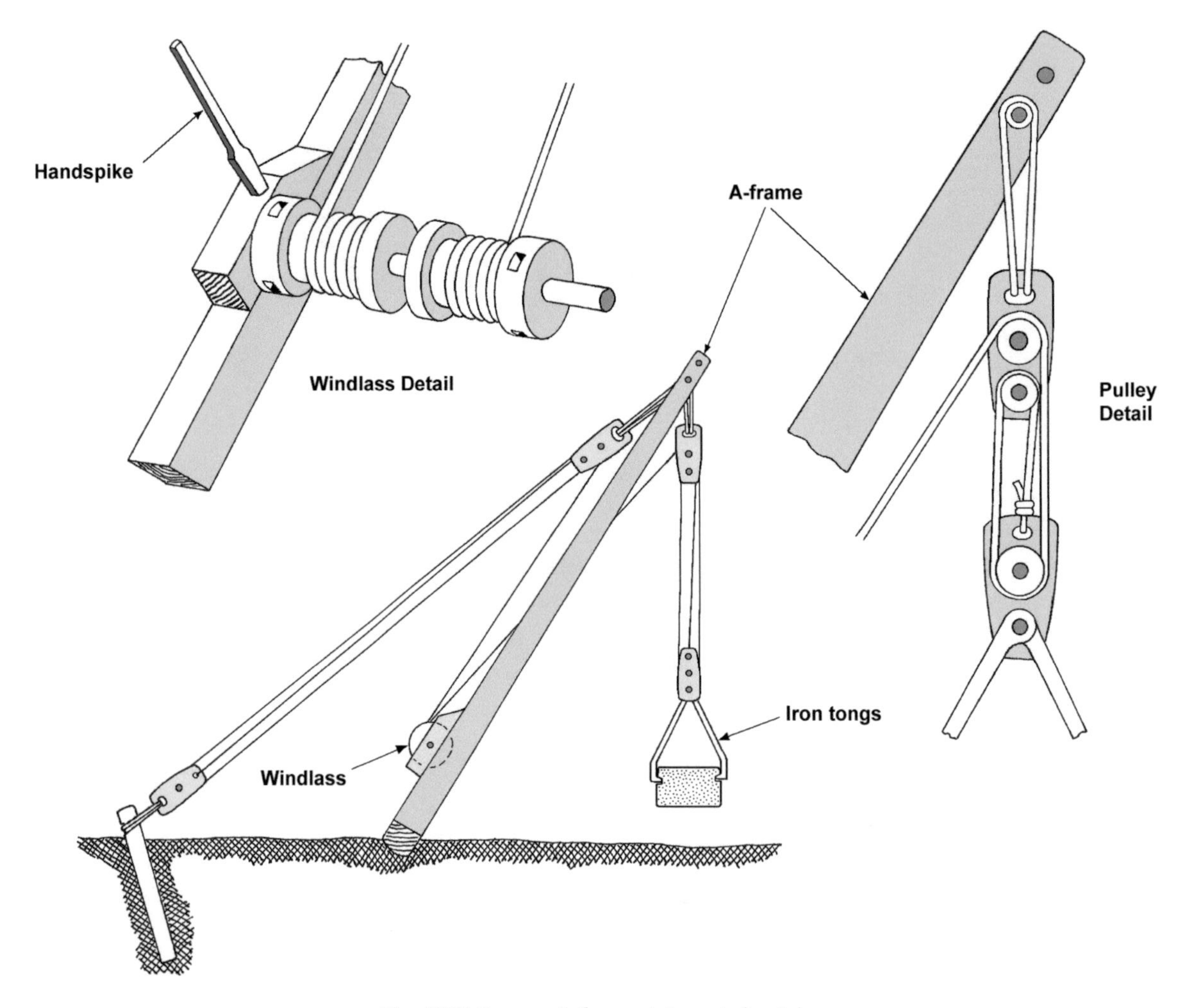

Fig. 225 Roman A-frame trispast derrick.

The *polyspast* featured pulleys with multiple sheaves on the same axle. A relief caving in the tomb of Qunitus Haterius (c. 100 CE) shows what appears to be an A-frame derrick with lifting force generated by means of a man-powered squirrel-cage instead of a handspike-levered windlass.

Some have proposed that the pyramid builders used similar hoisting machines. But we have never seen sheaved pulleys (or derricks) in ancient Egyptian paintings, carvings, or on models. This, and considering that the pyramids were raised 2,500 years before the Roman era, makes this idea implausible.

Lever-jacking

In the 1985 *Journal of the American Research Center in Egypt* (*JARCE*), Martin Isler proposed in his paper "On Pyramid Building" that the blocks were raised with levers, employing cribbing, up stone stairways that extended from each face of the pyramid. The small steps of the stairways would keep the cribbing stacks relatively short and stable.

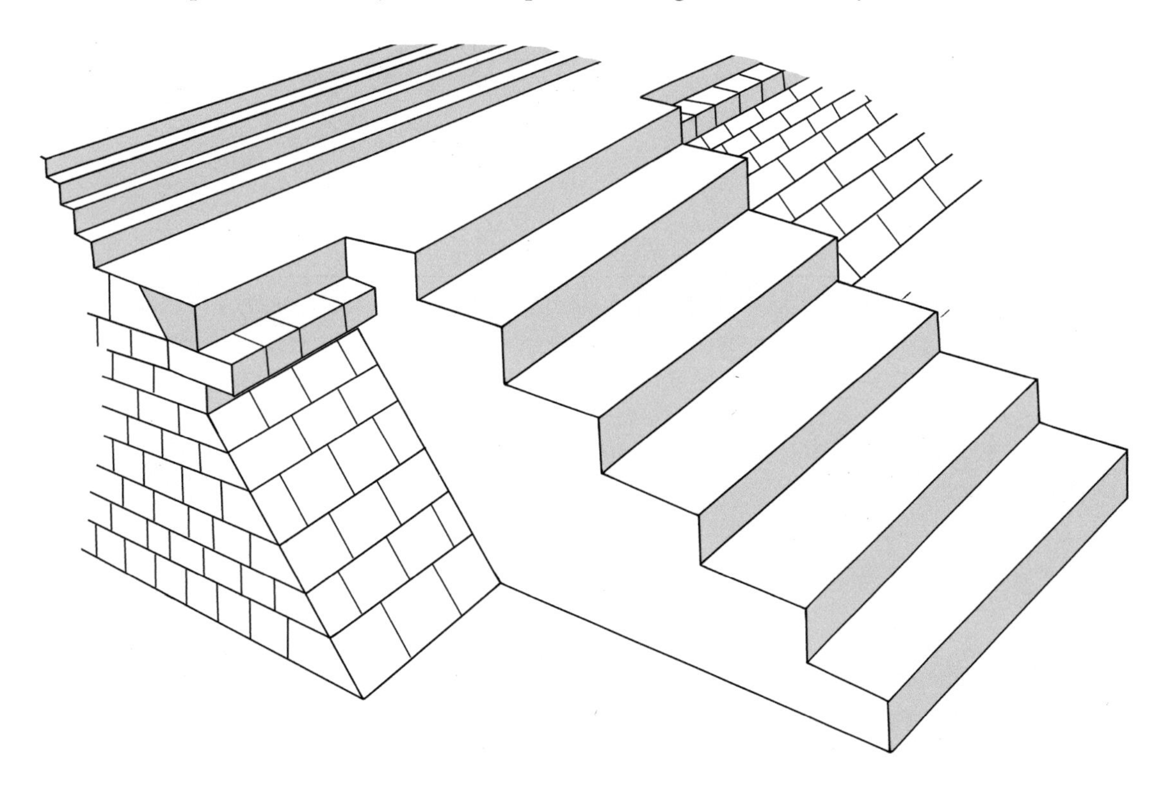

Fig. 226 The lever-jacking method envisioned by Martin Isler.

The stone-raising process would be the same for both stepped and regular pyramids. In another *JARCE* article (1987), Isler revised his proposal, saying that stone stairways were only needed in constructing stepped pyramids and stepped nuclei of regular pyramids. The filling of the steps in creating the final, regular pyramid form was done by jacking blocks over the casing stones which were placed in squared form before the final dressing smoothed them to a flat plane. English master builder Peter Hodges, working independently of Isler, offered a similar system in *How the Pyramids Were Built* (1989).

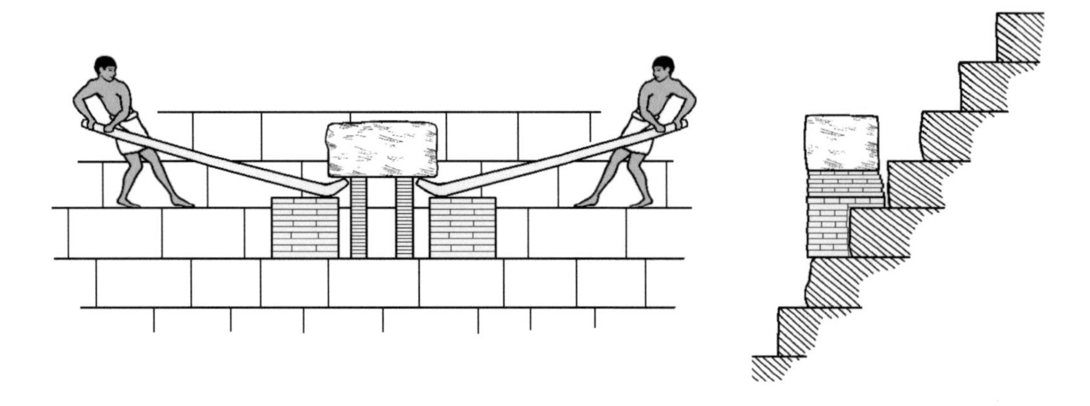

Fig. 227 Peter Hodges' lever-jacking method. Redrawn by author from Hodges (1989, fig. 24).

The strong points of Isler/Hodges are these: First, it is simple as machine theories go, employing small wooden levers and short wooden planks for cribbing. Second, we shouldn't expect to find remains of the simple tools because they would have eventually been used for other purposes. Third, lever-jacking allows many paths for taking stones up the pyramid sides, a time-saving benefit particularly advantageous for a relatively slow method.

The weak points of lever-jacking: First, and most important, the pyramids are too steep to take blocks safely up their sides by lever jacking. Tall stacks of cribbing would not be sturdy enough to support the heavy blocks, nor would they be well balanced while levers were being repositioned for another lift. On Khufu's pyramid the extreme variability in block size and shape would often result in situations where stones would be more than three times as wide as the steps over which they were being raised. This would require cribbing stacks to bridge at least two courses, and in some cases up to four, a recipe for disaster. This problem would be alleviated if Isler's additional stone staircases were used, which makes the idea at least feasible.

Second, lever-jacking, with its auxiliary work in building the extended stairways and placing/replacing the cribbing, is a slow process. Hodges found by experiment that an average-size block could be lever-jacked upward at 0.2 feet/minute (1989, 27). Conversely, dragging a sled-laden block up a 15 degree slope, by my experiments, yields a vertical speed for the block of eight feet per minute, forty times faster than lever-jacking, but requires ten times as many workmen. The speed handicap for lever-jacking can be partly offset because it can employ four times as many "tracks" as a big ramp. The delivery rate of stones to the working level of the pyramid then becomes about a tenth as fast for lever-jacking compared to the ramp. However, the lever-jacking speed deficiency would be worse still if we consider the time required to move each stone horizontally onto its new step.

Third, there is the problem of raising the 70-ton granite beams that form the ceiling structure of the King's Chamber in Khufu's pyramid, a vertical height of more than fifty meters. Hodges' solution was to build stone staircases of lesser slope within the body of the pyramid itself, but that does not mitigate the problem of instability inherent in stacks of wooden cribbing. An improbable means of lever-jacking the granite roof beams would be the scenario shown in figure 228, although no one, as far as I know, has dared to propose it.

Fig. 228 A possible, but unlikely, means of raising the roofing beams of the King's Chamber in G1.

Fourth, this idea, to be in contention at all, requires blocks to be delivered to the bases of stone stairways extending from all four faces of the pyramid. Considering that the quarries from which the stones originate are, in most cases, in one specific direction from the pyramids they serve, the stones would have to be dragged twice as far to get them to the stairway on the far side of the pyramid as to the stairway on the side closest to the quarry. That would seem wasteful of time and energy.

Lastly, we see no evidence of stone staircases either on or built within unfinished pyramids. The only pyramid that retains a substantial height of unfinished casing is that of Menkaure, but its blocks were not left in the squared aspect necessary for supporting stacks of cribbing or external stairways.

HOISTING

Tellefsen's direct hoisting

In the November 1970 issue of *Natural History*, engineer Olaf Tellefsen proposed that stones were hauled up greased wooden "skidways" by gangs of twenty-five to thirty men working from the top of the pyramid.

Fig. 229 Tellefsen's pyramid-side ramps.

At the ends of each skidway a "weight-arm" machine was used to lift the block onto or off of its skid-riding sled. Where the hauling ropes changed direction at the top of the pyramid, Tellefsen envisioned leather-wrapped greased logs as bearings.

Fig. 230 Tellefsen's weight-arm machine.

Tellefsen's vision of blocks being drawn straight up the pyramid side is attractively simple. If we omit his weight-arm machines, devices I believe unnecessary, this means for raising the stones would require the least contriving of equipment or structures for which there is no evidence. His wooden tracks are structures, but are so easily removed, and valuable for reuse, that we shouldn't expect to find them at the unfinished pyramids.

Another advantage of Tellefsen's method is that the four hip lines of the pyramid remain visible at all stages of construction (as did Isler), allowing continual accuracy checks.

A huge benefit of skidway hoisting is speed. Tellefsen's method not only allows many hoisting tracks, but is inherently fast. I figure the vertical speed of the block would be about 16 feet per minute, 80 times faster than Hodges' levering. However, Hassan's granite bearings would have been more efficient than leather-wrapped greased logs (resulting in greased ropes?) in reducing friction of the ropes passing over the top edge of the pyramid.

But how efficient are Hassan's bearings?

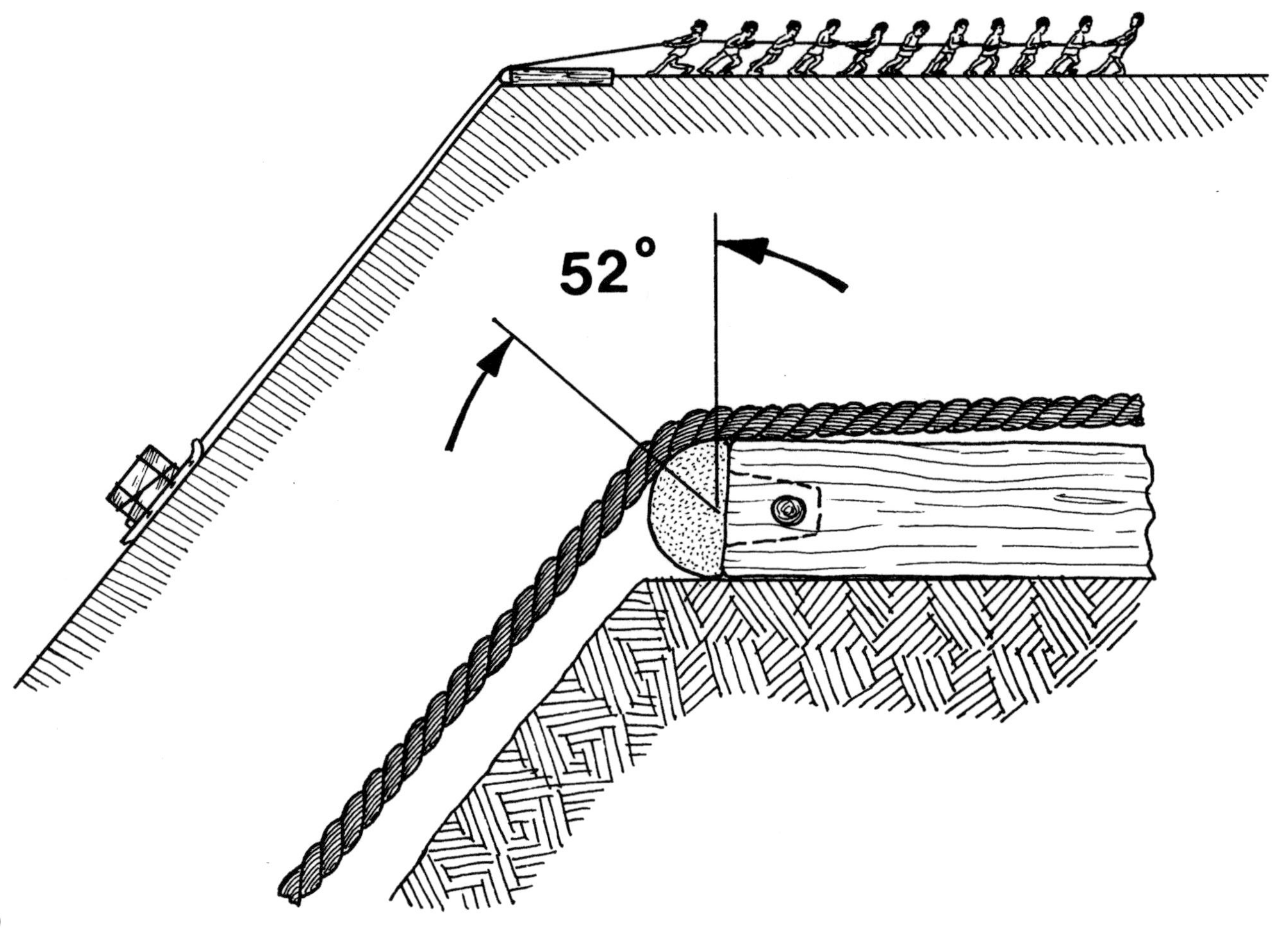

Fig. 231 Use of Hassan's granite bearing with Tellefsen's theory.

To evaluate the energy (work) lost to friction in Hassan's bearing one can use the *capstan equation*, $P = L \times e^{\mu ß}$, where **L** is the load to be lifted, **P** the minimum force required to keep the load moving at constant speed, **μ** the sliding coefficient of friction of the rope on the cylinder, **ß** the rope-to-cylinder contact angle in radians, and **e** the base of natural logarithms (approximately 2.7183).

I used hanging weights for L and P as illustrated in figure 232, and found the friction coefficient for dry hemp rope on polished granite to be 0.18. Wetting the rope increases friction, so water is no help.

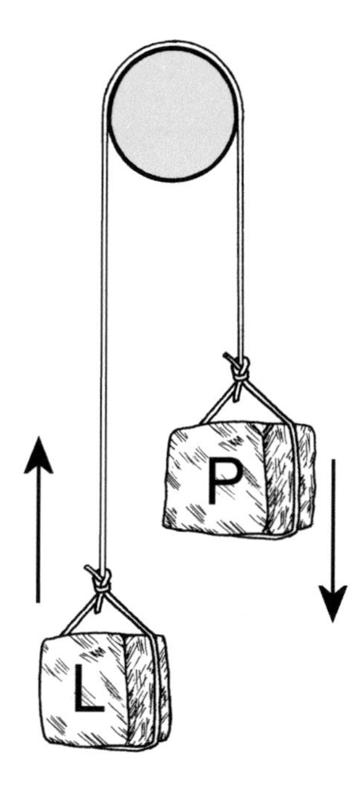

Fig. 232 Finding rope-to-cylinder friction loss.

For Tellefsen's idea to work with Hassan's bearings, the bearings would have to be elevated such that the sled-mounted blocks could be lifted beyond the corner of the course over which they were being hauled.

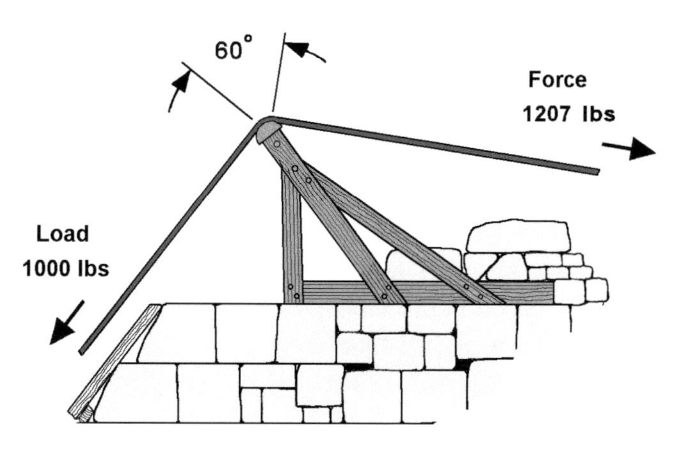

Fig. 233 Elevating Hassan's bearing.

With this arrangement a force of 1207 pounds would be required for every 1000 pounds of force on the load being hauled up the fifty-two-degree incline, an efficiency loss of seventeen percent at the bearing. The energy lost between the sled runners and the skidway is about three percent. The total loss to friction, then, about twenty percent of the effort expended, is not a high price to pay for the speed and minimal materials required to make the idea work.

Although Tellefsen did not address the building of stepped pyramids and stepped nuclei of regular pyramids, his idea would work well for those structures, especially if they were made of the smaller blocks in pyramids erected prior to the Bent Pyramid. In Djoser's pyramid average block weight is about 370 kg (816 pounds). It increased to roughly 545 kg (1200 pounds) in Sekhemkhet's pyramid, and to 770 kg (1700 pounds) at Meidum. For these buildings, Tellefsen's skidways would not be needed because the blocks could be hoisted up the near-vertical step faces (excepting Meidum E3), as shown in figure 234.

In chapter five I promised to speculate on the origin of the broad shallow grooves on eastern faces of the Meidum E1 and E2 projects. Perhaps these were places where blocks were hauled up the faces, creating deep scratches that were later removed by depressing the surface of the casing where it was damaged.

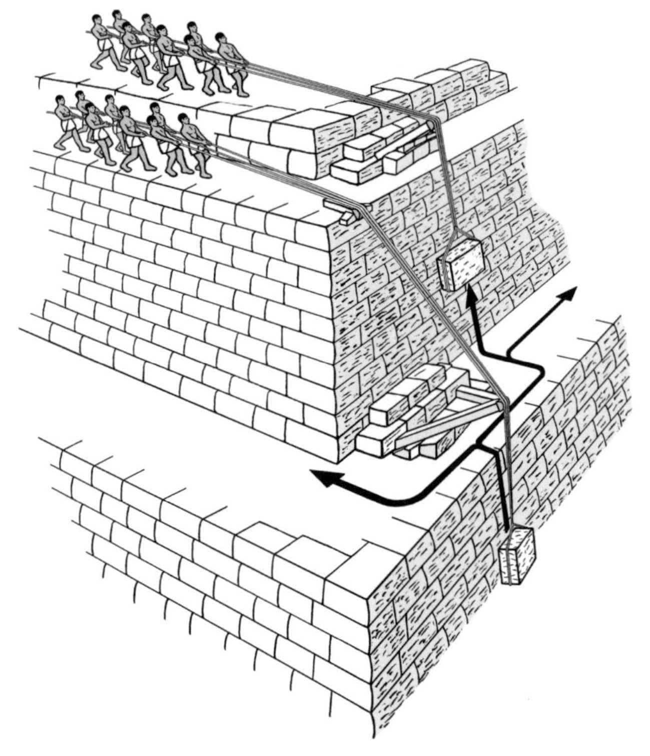

Fig. 234 Hoisting blocks up the faces of a step pyramid.

Tellefsen's theory, however, has flaws. On the unfinished granite casing of Menkaure's pyramid we see no evidence by which skidways would have been anchored. Nor do we see scratch marks on the casing blocks if skidways were not used.

The ropes needed to hoist blocks from ground level to the top of the pyramid become unacceptably long. For Khufu's pyramid, where the apothem is 186 meters (610 feet), we must add at least another thirty meters for the haulers, making the ropes, at minimum, 216 meters (710 feet) long. Putting aside the maintenance required to keep a rope of this length in a continuous piece, the rope itself would become a substantial (and wasted) portion of the load to be lifted.

Another problem is heat in the stone bearings produced by friction of the ropes sliding over them. The energy lost to friction at the bearing, as I mentioned above, is about 17 percent of the total energy expended, and is exhibited as heat. Let's say the rope absorbs half of this energy and the bearing the other half. Heat in the rope is not a problem because it is spread out along the air-cooled rope. Not so for the bearing. Heat is concentrated there in a small mass with poor heat dissipating ability. The bearing would have to be constantly doused with water to keep it from overheating and burning the rope. This issue does not alone defeat the theory, but must be accounted for.

The biggest drawback for Tellefsen's idea is that it cannot accommodate the granite ceiling beams of Khufu's King's Chamber. Why is this so? Let us again use the capstan equation. The force needed to pull a 75-ton block strapped to a 2-ton sled along the inclined face of Khufu's pyramid is 150,000 pounds. The force required of the rope haulers on the top surface becomes 181,000 pounds. If each workman was able to pull with a force of 70 pounds it would take nearly *twenty-six hundred* workmen to raise the largest of the King's Chamber roof beams. This would seem to create a space problem for the haulers.

Though I doubt the King's Chamber roof beams could have been raised by hoisting, I do not rule out hoisting for smaller blocks, especially those near the top of the pyramid. I'll say more about this when I offer my idea later in this chapter.

In *Sticks, Stones, and Shadows, Building the Egyptian Pyramids*, Isler revised his stone-raising theory again. He kept his idea that stepped nuclei were erected by levering/cribbing, but to fill the steps needed for the true pyramid form he presented a concept similar to that of Tellefsen: Stones were hoisted up the sides of the pyramid by teams of men standing on steps of casing stones which had not been finish-planed to the final pyramid slope. To allow the stone-haulers periods of rest, rope forces were momentarily relieved by a "snubbing device."

One problem with this idea is that the snubbing device, as shown, is not anchored sufficiently to the steep pyramid face. It would not hold the force of the weight without tipping over. In figure 235, the moment (or torque) M about point O would cause the snubbing device to rotate away from the pyramid face if not anchored somehow. Even if the anchoring means included parts of casing blocks that would be later trimmed away, and thus leave no evidence, it is difficult to imagine how this system could be made reliable enough to do the job Isler proposes. Another problem is similar to the first: the workmen are not anchored to the steep pyramid face. In hauling on the ropes, or employing the levers he pictures, it looks to me like they could not safely stay on the pyramid steps.

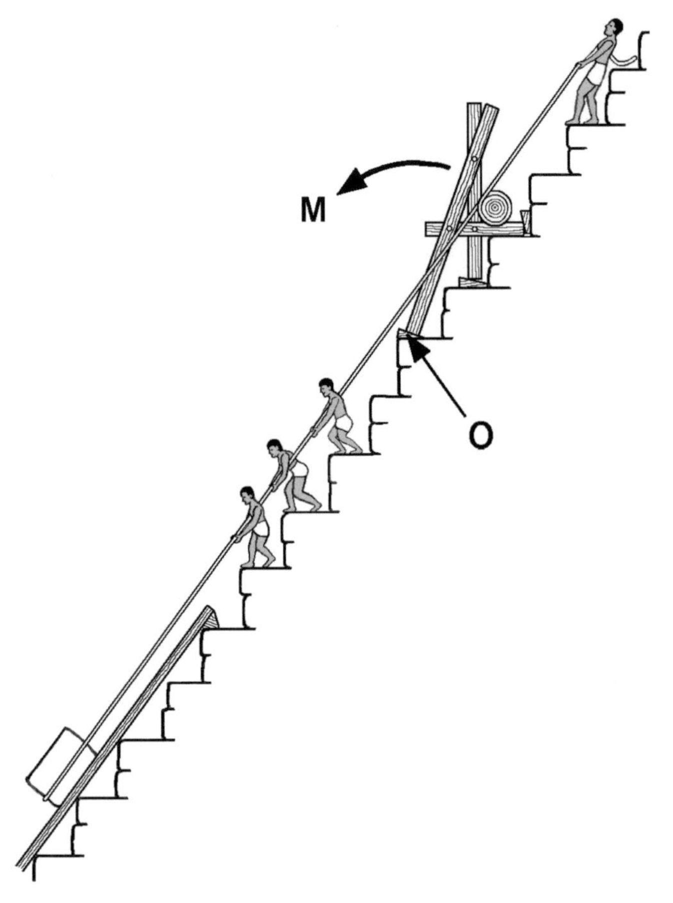

Fig. 235 Isler's hoisting method using a snubbing device. After Isler (2001, fig. 11.13).

At the top of the pyramid, Isler suggested that workmen dragged the blocks up and over the edge using a wooden frame in which Hassan's stone bearings were fixed. This arrangement would lessen the frictional loss of ropes running over wooden poles or the block edges themselves. I agree that this is a good idea, but Isler envisions two stone bearings where only one is needed. With two changes of direction of the hauling rope the friction loss is almost twice that with one change of direction. Referring to figure 236, Isler's method would require, for example, a 1706 pound-force to generate a 1000 pound hauling force, a loss of forty-one percent compared to seventeen percent using the single stone bearing arrangement illustrated in figure 233.

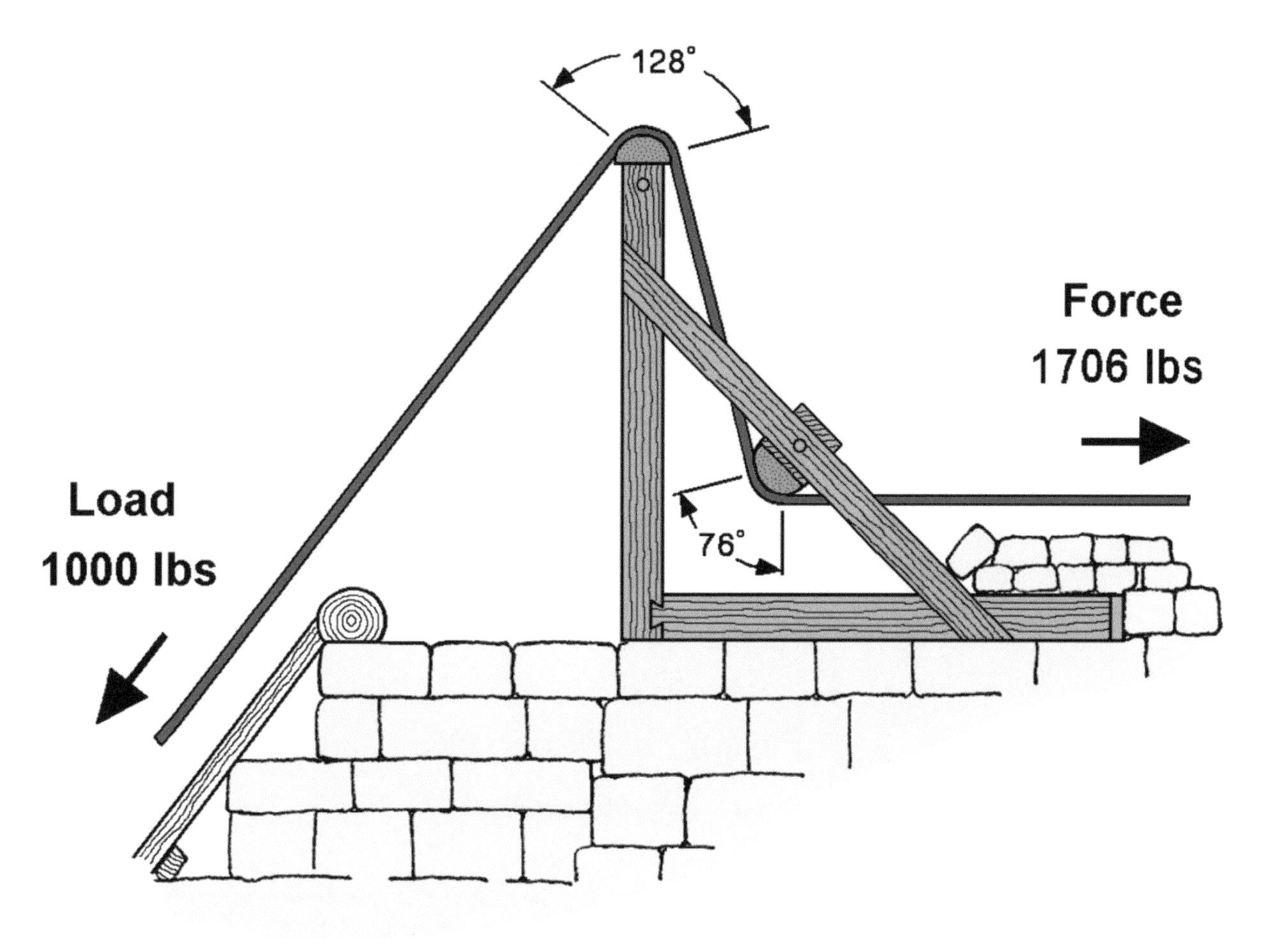

Fig. 236 Isler's double-bearing arrangement. After Isler (2001, fig. 11.17).

Finally, this idea has the same drawbacks as Tellefsen's: the requirement of very long ropes and, even with the slightly-less-steep staircases he proposes, the inability to raise the largest blocks in G1.

Lohner's Rope Roll

In 2006 German engineer Franz Lohner published on the internet his idea of how the Great Pyramid was built. Lohner employs a hoisting method using what he calls "rope roll" devices (which I will call RRD's). The RRD's are akin to a pulley system whereby workmen can walk down wooden ladders on the face of the pyramid while hoisting the stones upward.

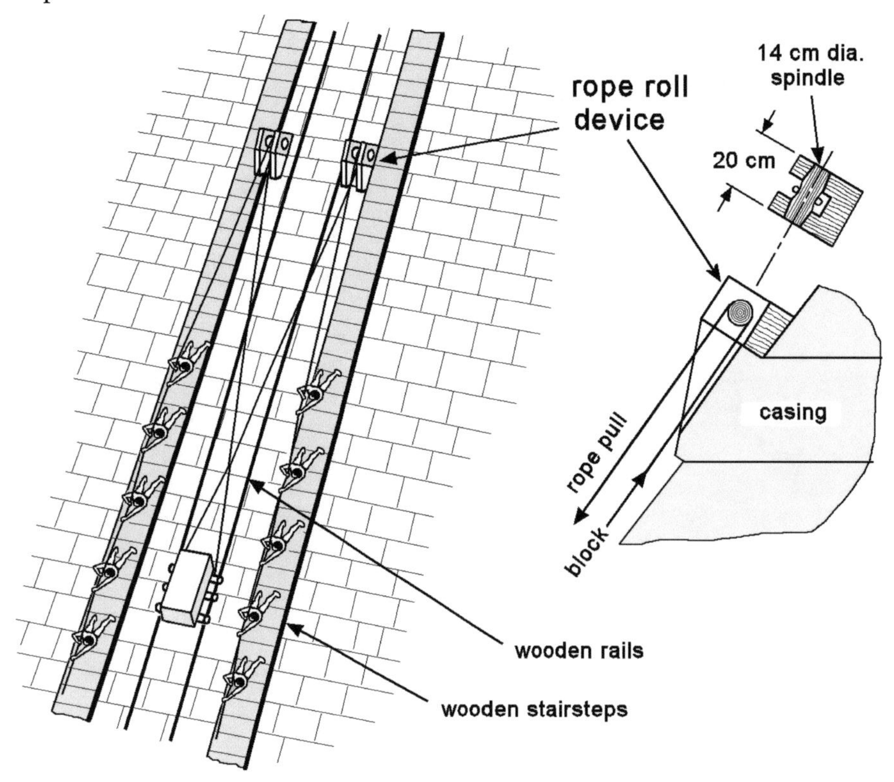

Fig. 237 Lohner's "rope roll" hoisting system. The two RRD's are enlarged for clarity.

The roller in Lohner's device is a copper-sheathed wooden spindle captured in a wooden cradle containing copper sleeves which support the ends of the spindle. In this concept the RRD's are located on the face of the pyramid, 30-37 meters vertically apart, so the blocks can be transferred from one device to the next as they are hoisted up the pyramid face. The workmen walking down the stairways are pushing cross-beams (not shown in my drawing) to which the hauling ropes are attached. The block being hauled is tied to a sled having the form of a framework of logs. The sled rides on wooden rails fastened to the pyramid faces. The RRD's are supported by special casing stones having protrusions to which the devices are lashed.

Lohner's proposal is good in principle. Workmen walking down the pyramid face while pulling blocks up by means of a pulley system anchored above is a great idea.

Multiple tracks and fast-moving blocks equates to fast construction. As in Tellefsen and Isler, the four hip lines of the pyramid are always exposed, so maintaining accuracy of the building is not a problem.

Also to Lohner's credit is that he built and tested a model of the system, including one full-scale test in which he hauled a 5-ton block up an incline using a fork-lift as pulling power. He does not state how far the block was hauled.

Now to the problems with Lohner's idea.

First, there is a strength-of-materials issue with his RRD's. Wood makes a poor axle for a heavy lift machine, and copper is not a good bearing material for the same reason, especially copper surfaces moving against each other. The wooden spindles, 14 cm (5.5 inches) in diameter would deflect under the multi-ton forces we are talking about. The deflection would put cross-corner loads on the copper bearing sleeves which, not being suitable for bearings in the first place, would rapidly wear, if not weld together. Continuous lubrication, which Lohner admits would be necessary, would delay the inevitable for only a short time. Lohner says that due to wear of the ropes, spindle, and bearing sleeves these components would have "to be replaced often." This is an understatement. Here one cannot extrapolate conclusions from a small model to a full-size machine.

Second, I doubt the cantilevered protrusions he shows on the casing blocks would be either strong enough or last long enough to handle the many thousands of blocks being raised on each track. And if any portion of the two projections per RRD breaks off, that path up the pyramid would be taken out of play.

Third, the system he describes for raising the granite monoliths for Khufu's crypt ceilings is complicated. It involves huge counterweight blocks that have to be first hauled up with RRD's so that they can provide the force, sliding back down, required to raise the granite roof beams.

Last, there is a striking lack of evidence for the system he envisions. No devices or parts of devices he imagines have ever been unearthed in Egypt. No casing stones on the pyramids with intact casing show signs of anchored track-ways or stairways. There are no features on the granite casing of Menkaure's pyramid that would suggest the use of RRD's, stairways, or tracks.

Weakening all machine theories are unidirectional quarry locations. If machines were used to raise the stones why wouldn't the stones have been obtained from all directions around the pyramid? In this way, transport distances could have been greatly reduced. This objection could also be pointed at certain ramp theories, especially those which envision ramps approaching the pyramids from all directions.

RAMPS

Ramp theories are more numerous, but all are variations or combinations of three types: switchback, straight, and spiral. All envision sled-borne stones being dragged along by human power.

The Switchback Ramp

W. W. Lucker imagined a ramp comprised of a series of switchbacks on one side of a stepped core.

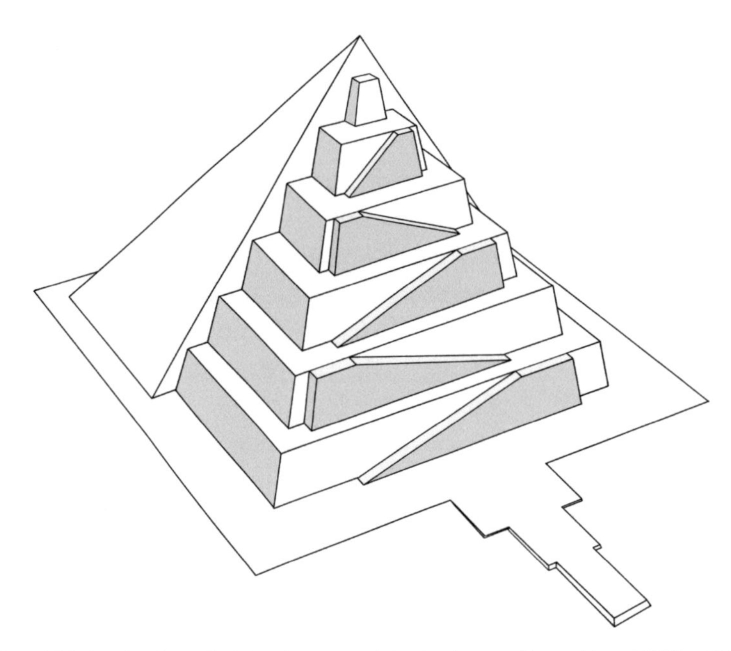

Fig. 238 Lucker's switchback ramp. Adapted from Tompkins (1971, 225).

There are many problems with this idea, but two are fatal indeed. The ramps near the top become not only too steep for block-dragging, but also overlap, a geometric impossibility. Second, a ramp *within* the body of a pyramid cannot be used to cover itself (proceeding top to bottom) in creating the true pyramid form unless the law of gravity can be temporarily repealed. If these issues were overcome by disposing the switchback ramp *outside* of the final pyramid envelope, its volume (and work) would have exceeded that of the pyramid, a situation I doubt the builders would have accepted.

The Big Straight Ramp

Borchardt and Lauer suggested sled-borne stones were dragged up a large straight ramp built against one face of the pyramid.

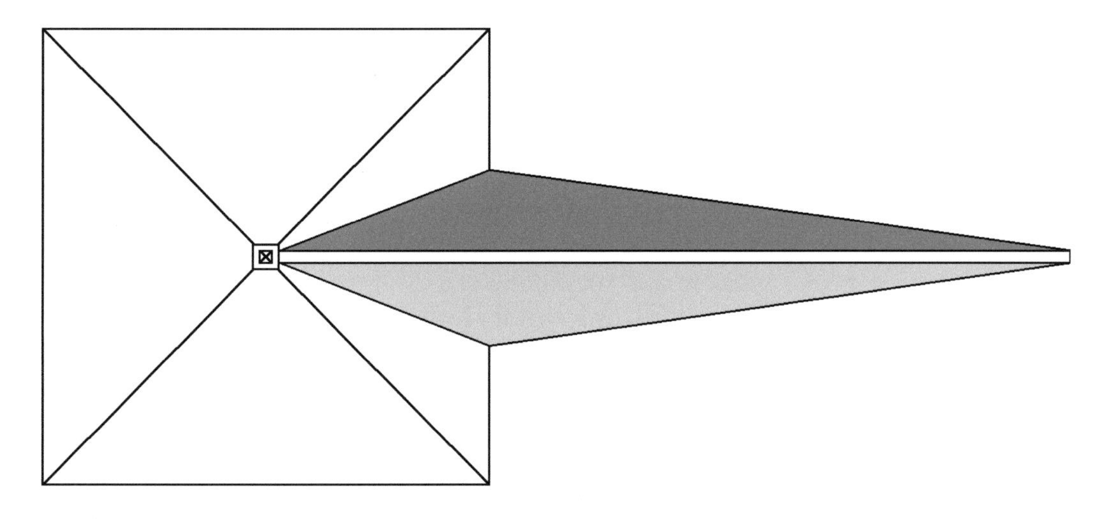

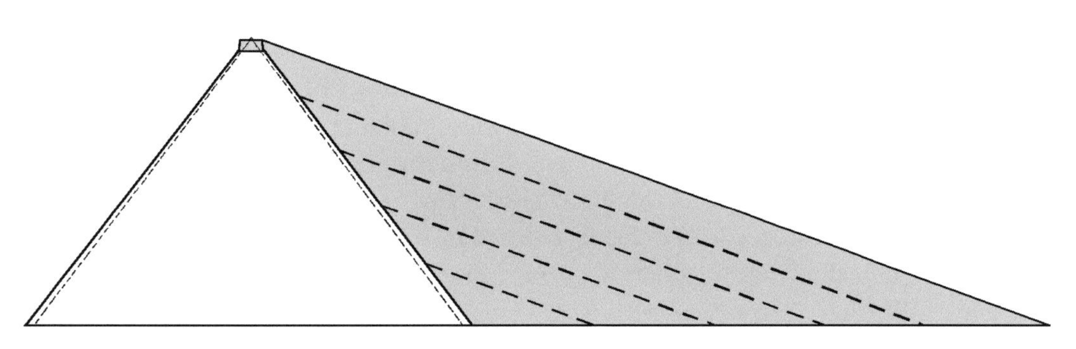

Fig. 239 The straight ramp. Adapted from Lauer (1974, fig. 62).

The angle of this ramp is kept constant. As it is extended upward, it is also extended outward. The ramp has rather steep sides, about 68 degrees, given the geometry they show. It eventually reaches the top of the pyramid where there is a working platform 16 meters square (Lauer). Lauer's drawings show a coating of brickwork on the outside of the pyramid such that the casing stones could be pushed into position from these embankments, as proposed by Clarke & Engelbach.

Clarke & Engelbach did not offer a hypothesis on the configuration of the supply ramps, saying only that "in the case of moderate sized blocks, they may have been rolled up the slopes, and only mounted on sleds when they had to be moved any considerable distance horizontally." How a multi-ton block could be rolled up a slope, Clarke & Engelbach did not say.

I.E.S. Edwards proposed (1972, 206) a ramp similar to that of Borchardt and Lauer. Edwards' ramp has sides that slope less steeply, at about 45 degrees, but is more voluminous. He also liked the idea of foothold embankments surrounding the entire pyramid.

The straight ramp has advantages in being simpler than switchback and spiral ramps. No time-consuming turns are needed and it rests well supported against one face of the pyramid. On the other hand, it has several weaknesses. Borchardt and Lauer depict a configuration where the ramp rises constantly at a 20 degree angle for all stages of building the pyramid. I believe this is too steep for general sled-dragging, though in the last phase, for a small portion of stones, it may have been acceptable. Further, the sides of their ramp slope at an angle steep enough that its sides would not be self-supporting unless faced with mortared stone blocks. In that case it does not seem reasonable that on such a labor intensive project as a pyramid, the builders would construct an equally labor intensive auxiliary work.

Another problem with the big straight ramp that goes all the way to the top of the pyramid (I'll explain later why this does not have to be so) is its incompatibility with the building of a nucleus structure in advance of the final pyramid shape. At Meidum for certain, and probably at other sites, stepped nuclei were either enlarged in separate projects, or at least advanced somewhat beyond the addition of the casing and filling that formed the true pyramid shape. In the case of the Meidum pyramid, it seems unlikely that a huge ramp was completely built and partially or wholly dismantled *three* times.

Last, if the pyramid was completely covered during construction, as proposed by Lauer, Edwards, and Clarke/Engelbach, the hip lines could not have been observed during construction. Thus, edge straightness would not have been assured. I do not think the architects would have accepted that risk.

The Interior Ramp

As I mentioned on Hodges' lever-jacking method, he proposed that the granite roofing beams in G1 were lever-jacked up stone stairways built in the body of the pyramid. In *Building in Egypt* (1991), Dieter Arnold, Curator in the Egyptian Department of the

Metropolitan Museum of Art in New York, took that idea further to include internal ramps still higher(see figure 240).

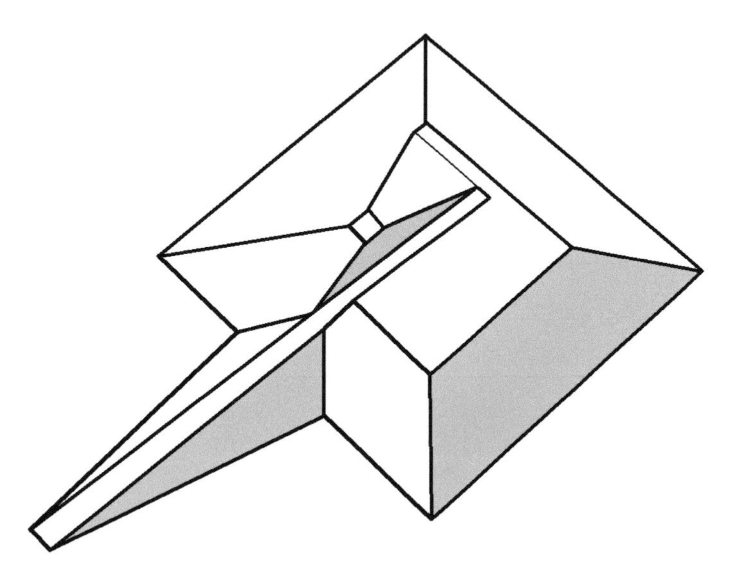

Fig. 240 The interior ramp. Adapted from Arnold (1991, fig. 3.53 [4]).

Supplementing the interior ramp, Arnold says (1991, 101) "Since these open trenches would have been filled in after some time, an added exterior ramp would have been necessary. Not much material would have been required, so this exterior ramp could have been comparatively small, perhaps in the shape of a staircase ramp." Of the staircase ramp Arnold says "Naturally, the building stones could not be pulled up the steps but would have been lifted with levers or other devices." He offers no details on what these other devices might have been, nor any evidence for staircase ramps, interior or exterior.

Arnold allows that "The disadvantage of this system is that the construction trenches would have disturbed the proper building of the core steps and accretion layers. Nevertheless, construction gaps, probably from such interior ramps, could be seen in the core masonry of the pyramids of Sahura, Niuserra, Neferirkara, and Pepi II."

Given that construction gaps are found in several unfinished pyramids, I have seen no evidence that these gaps served a purpose beyond the building of internal apartments (typically burial chambers) at lower elevations of the buildings. They may have allowed erection of core masonry *independently,* at least for a while, of the work on the crypt and its access shaft.

The main problem with Arnold's idea: If you look carefully at the drawing he offers (as in figure 240 above), it is difficult to see how the interior ramps could be extended to build the top of the pyramid. Thus, his idea does not meet the second of my twelve criteria which any stone raising theory must satisfy. To Arnold's credit, however, he does not insist, as many theorists do, that the method he envisions was, in fact, used.

The Exterior Spiral Ramp

English scholar Moses Cotsworth (1859-1943) proposed the spiral ramp theory, where four ten-foot-wide earthen ramps spiral up and around the faces of the pyramid. Three of the ramps were employed for hauling stones upward, the fourth for return of empty sleds to the quarry.

Fig. 241 The spiral ramp theory.

This configuration looks superior to the straight ramp in that much less ramp volume is required. However, its deficiencies are many.

First, the nearly vertical sides of the ramps are too steep for earth, clay, or even mud-brick; they would not stick to the pyramid. This by itself is not a huge drawback. Substitute stone masonry as shown in figure 242 and that problem goes away.

Fig. 242 Cross-section of spiral ramp.

Second, like switchback and interior ramps, this system cannot get blocks to the top of the pyramid. Near the top these ramps either overlap or become impracticably steep.

Third, like switchback ramps, the sleds have to be turned many times as they negotiate the corners of the building, substantially hindering work flow.

Fourth, it is difficult to envision how the 70-ton granite roof beams in G1 were hauled up and around the sharp bends of the ramps.

Fifth, as the building rises, the hip lines cannot be line-sighted or built to a stretched cord. Last, no evidence for spiral ramps can be seen at the unfinished pyramid sites.

The Integral Spiral Ramp

Dietrich and Rosemarie Klemm (2010), proposed a pair of spiral ramps integral with the core masonry of the pyramid. One ramp would allow dragging stones upward, the other for return of empty sleds to the base.

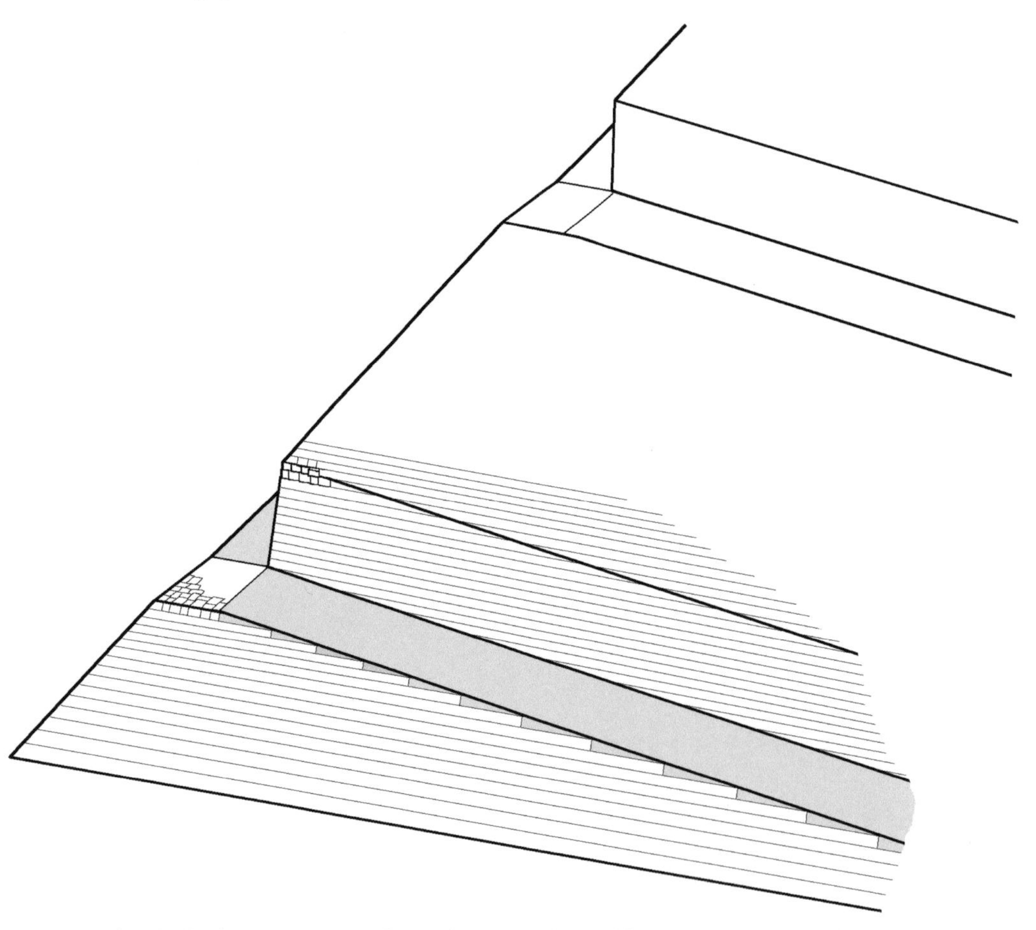

Fig. 243 Klemm integrated spiral ramp. Adapted from Klemm (2010, fig. 85).

One advantage of this ramp is that no ramp-constructing materials are needed. Another is the hip lines are always visible.

Though this idea seems ingenious, it has numerous problems:

1. The pyramids have leveled courses for good reason. It allowed periodic checking of the square as the buildings rose. If the courses are substantially interrupted the square cannot be measured.
2. Turning the corners, as with other spiral and switchback ramp ideas, would not be easy or fast. There is no run-off for the line of block-haulers that would allow them to drag the block up to the corner of the pyramid where it would have to be turned ninety degrees. This problem would be exacerbated in raising the roof beams of the King's Chamber in G1.
3. Spiral ramps within the body of the pyramid create at the top of the pyramid an even greater geometry problem (if there could be increments of impossibility) than exterior spiral ramps. Build a model and you will see.
4. If integral ramps were used we should see, I believe, indications in the backing stones where these ramps were filled in, but no such evidence appears on any pyramid.
5. Lastly, how were the ramps filled in? They couldn't be filled in from bottom to top because the filling would immediately remove the means of transporting more filling blocks upward. If filled in from top to bottom it is hard to imagine how this could be done without an anti-gravity machine, since it would entail, for most of the filling blocks, inserting a lower block beneath one above. Think of building a brick wall starting at the top and working your way down!

The Interior Spiral Ramp

In July 2006 French architect Jean-Pierre Houdin (1951 – present) published *Khufu: The Secrets Behind the Building of the Great Pyramid,* a wonderfully illustrated book containing an ingenious, intricate thesis which I will attempt to summarize. I devote more time to this idea because it seems reasonable to laypersons.

Houdin proposed that the Great Pyramid was raised using several ramp configurations. First, sled-riding stones were dragged to a height of 43 meters (141 feet) via a wide straight ramp exterior to the pyramid in the manner of Borchardt and Lauer. From that level a series of switchback ramps within the body of the pyramid itself were employed to raise a nucleus structure to the height of 76 meters (249 feet).

Simultaneously with these two phases, an *internal* spiral ramp, initially 2.6 meters wide, starting seven meters above zero level at the southeast corner of the pyramid, disposed three meters inside the outer casing, was constructed as the pyramid rose. Houdin attributed this key idea to his father Henri, whose theory Jean-Pierre set out to prove.

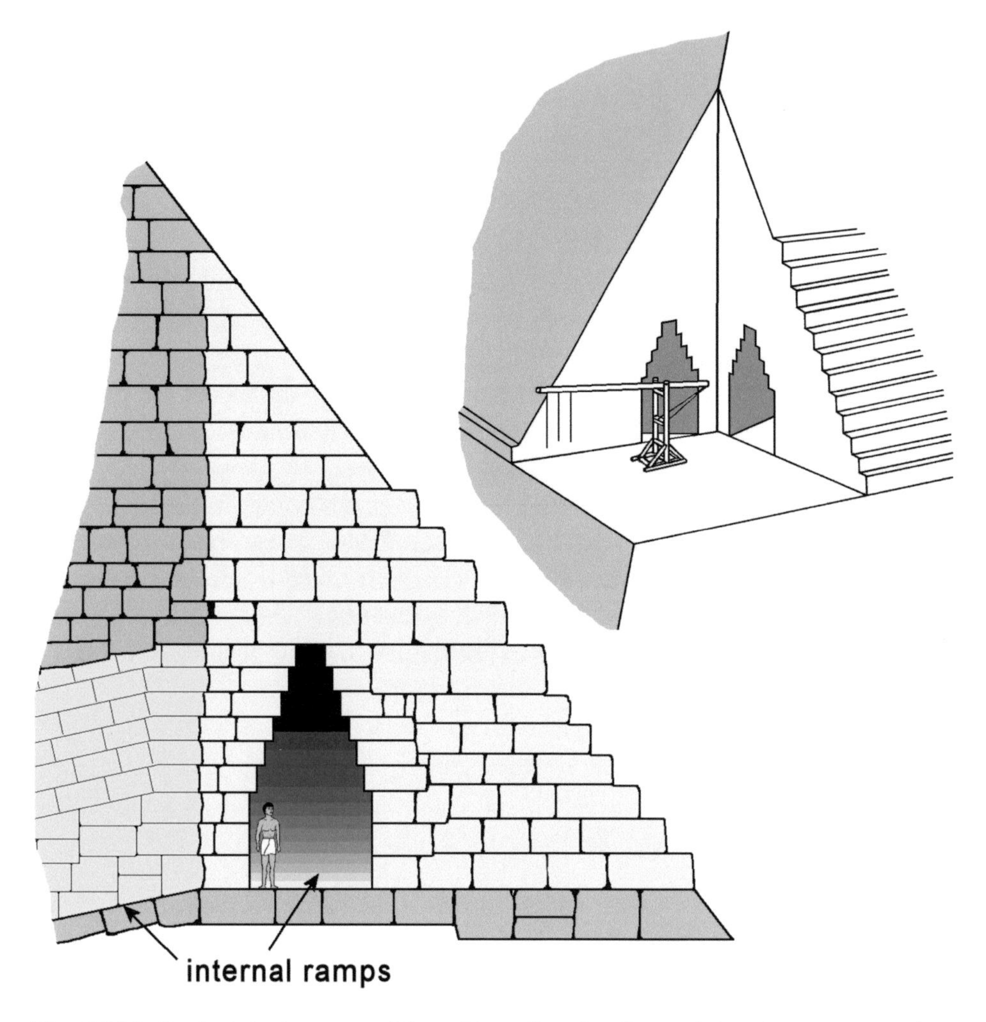

Fig. 244 Left: the interior ramp of Jean-Pierre Houdin. Upper right: the corner notches of the internal ramp, where shadoof-type machines were used to turn the blocks. Exterior wooden gangway not shown. Adapted from Houdin (2006, 66).

The internal spiral ramp was employed only after the first two stages were completed. To raise the blocks around the inner core and above, that is, from the 43 meter level upward, the inner ramp was used. The internal ramp needed be only 2.6 meters wide up to the 123 meter level because, says Houdin, the blocks transported to that level weighed "a ton maximum." Above the 123 meter level the width of the internal ramp was reduced to 2.1 meters because it had to accommodate blocks "weighing little more than half a ton." From the 130 meter level, blocks were hoisted the last sixteen-and-a-half meters by means of a "wooden lifting tower." Houdin does not tell us how it worked.

Along with the internal ramp, a two-meter-wide wooden ramp, or "gangway," affixed to the pyramid exterior (how affixed he does not say), was constructed parallel to the internal ramp. The external gangway was used to return "traction teams" (block draggers) to their last landing after hauling blocks up the particular straight section of the internal ramp to which they were assigned.

At each corner of the pyramid where the internal ramp changed direction was a large notch having a base 10.5 meters (34 feet) square, and open to the air. On these platforms "crane drivers" used shadoof-like machines (ala Croon) to lift and turn the sleds. Adjacent to each notch was an expanse of casing stones which had not been planed to the pyramid angle. Blocks which would eventually be used to fill in the notches were temporarily stored there. The internal ramps would never be filled in.

To raise the megaliths of the King's Chamber, Houdin proposed an elaborate counterweight system.

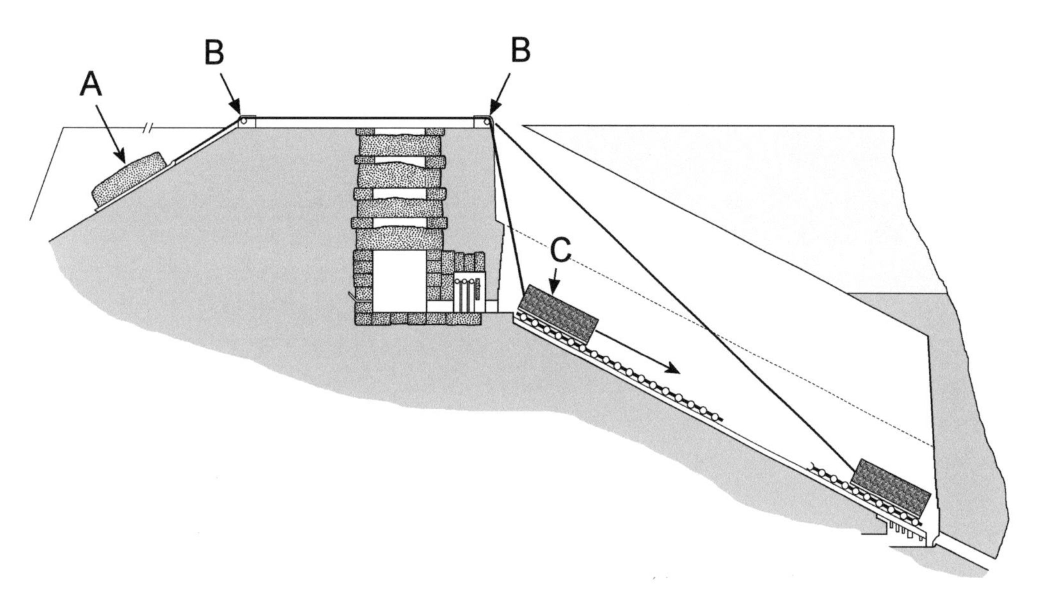

Fig. 245 The counterweight system used to raise the ceiling beams of the King's Chamber. Adapted from Houdin (2006, 85), with some details omitted for clarity.

A 25-ton counterweight (**C** in figure 245), traveled on a multi-roller trolley that rolled on top of the two benches of the Grand Gallery. As it moved down the Gallery it pulled ropes that ran over roller bearings at **B**, and down to the roofing beam **A** being hoisted. The ceiling of the Grand Gallery had to be twice as high as it is today to allow clearance for the counterweight ropes.

To reload the counterweight a force of eighty workmen disposed on the top level of the pyramid, using a pulley arrangement, pulled it back up the Grand Gallery.

Coordinating the raising of the King's Chamber rafters, an operations director, located at the upper terminus of the northern air duct of the Queen's Chamber, used that shaft as an "intercom" to communicate with his counterpart at the southern terminus of the southern air duct. By shouting directions down each shaft, and taking advantage of acoustic amplification of the Queen's chamber, messages were exchanged (notwithstanding the fact that neither director could have seen each other or what was going on in the Grand Gallery).

As the pyramid rose, the capstone was elevated along with it by means of a pyramidal wooden cage to which the stone was connected by four ropes. To begin each lift, workmen soaked the ropes with water and stretched them tight. The ropes were allowed to dry in the sun, thereby shrinking and lifting the stone enough that workmen could rotate it. Twisting the bundle of four ropes, in the manner of a Spanish windless, would shorten them, lifting the stone 10-15 cm per cycle. Block up the stone, then repeat the process.

Evidence Houdin cited in support of his internal ramp is an image from a microgravimetric survey conducted by a French team (EDF Foundation) in the 1980s. The image appears to show, in Houdin's opinion, a spiraling pattern of lesser density within the pyramid. He also noted the large notch about two thirds of the way up the northeast edge of the pyramid. He thought this place could be one of the block-turning locations.

In April 2008, American Egyptologist Bob Brier investigated the notch on behalf of Houdin, and reported details in *The Secret of the Great Pyramid* (2009). The book had the somewhat presumptuous sub-title *How One Man's Obsession Led to the Solution of Ancient Egypt's Greatest Mystery*. The notch has a base about five meters (16 feet) square, or one quarter the area Houdin hypothesizes for the block-turning platforms. At the inside corner of the notch is a crevice about a foot-and-a-half wide and five feet high. The crevice leads to a cavity which Brier called "the cave."

Later, Brier wrote in an article for *Archaeology* (2009) that the cave is an "L-shaped room," with legs of the L roughly 3.3 meters (11 feet) long. [I observe from Brier's photos, however, that the room appears more irregular-shaped than L-shaped]. Brier noted that some of the blocks in the walls are "carved in semi-arches to support the ceiling," and "the central ceiling block was set in place, like a keystone," the implication being that the cave is a purpose-built chamber. Also in the cave Brier found the number (date?)

"1845" painted on one of the blocks. Brier thinks the cave was built to facilitate filling of the notches when the internal ramp was no longer needed for pyramid construction.

Now to the problems with Houdin's theory.

Where do I begin?

I'll start with his physical evidence.

The microgravity image is puzzling. The bands of lower density appear mostly *parallel* to the pyramid's base edges, not in a spiral pattern. The bands widen in certain sections, expansions which could be imagined as partly spiral if that is what one wants to see. More importantly, the areas of lower density, if that is what they are, are 10-25 meters wide, substantially greater than the 2.6-meter-wide ramps Houdin envisions. Therefore the image does not support Houdin in any manner I can see.

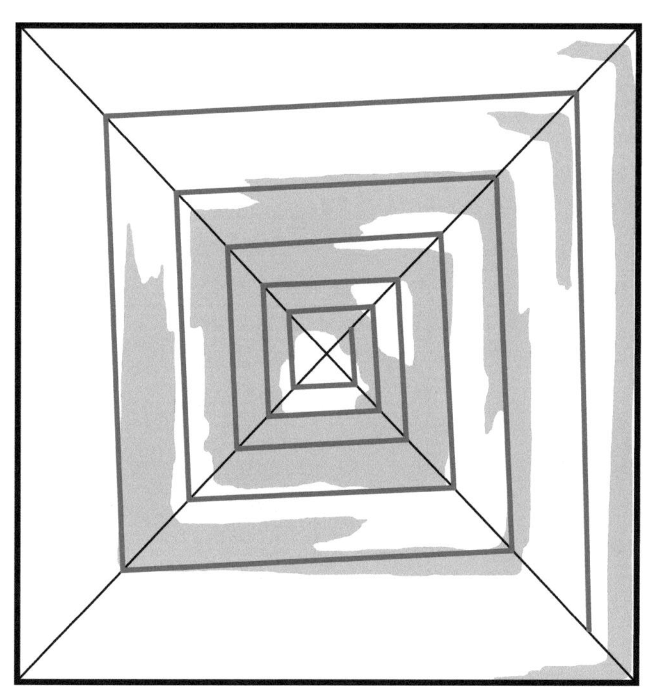

Fig. 246 The EDF microgravimetric image modified by the author in removing everything except the areas of lower density, in green, which Houdin suggested might be evidence of the internal ramp. Houdin's actual proposed ramp is shown as the dark blue line.

The notch and cave. How did the current notch, which is half the height hypothesized by Houdin, come to be? The builders wouldn't have left this notch unfilled, would they? It follows, then, that a later group created it. Who were they and why did they remove stones to create the notch? Perhaps casing-strippers in the Middle Ages made the notch and hollowed out the cave in order to make a resting place and temporary respite from the brutal summer sun.

The notch was a popular resting spot for climbers, at least until my last ascent in 1987. Greg and I briefly stopped there on our descent in 1978 to converse with an elderly Russian tourist who had taken refuge before continuing his climb. The crevice to the cave, I regret to admit, we overlooked. To Brier's point that large blocks of the cave could not have been taken out through the narrow crevice I offer that the largest blocks among the mixed sizes typical in interior masonry can be cut into smaller blocks. Further, his photographs do not indicate a purpose-built arch in the cave, only the carving away of protruding corners of blocks to prevent head-knocking. I see no evidence of the interior ramp in the cave or in the notch.

Additional problems:

There is no evidence of the internal ramp where it enters the pyramid at the southeast angle or where it exits the pyramid at the top. Significant stone has been removed from the southeast base of the pyramid, more than enough to reveal the imagined tunnel. No signs of the tunnel or plugging thereof exist. The tunnel *exit*, according to Houdin, would be at the 130 meter level, which is only 7.5 meters below the present top. I examined the top ten meters of the pyramid as I looked for machine evidence in the form of holes in blocks. And though I was not looking for small blocks filling the end of Houdin's ramp (or notch), I did not notice anything unusual in the masonry, certainly nothing to indicate the plugged exit of an internal ramp.

How could large blocks that lie above the 43-meter level be taken through tunnels only 2.6 meters wide? Houdin says the internal ramp "was to be set up only for the transport of small-sized blocks weighing a ton maximum." Even if he is using metric tons his weight estimate is way off. Many of the backing stones on the 118th course, 90 meters above zero level, are 0.9 meter thick and weigh at least six tons. The corner casing blocks probably weighed eight tons. As for his statement that blocks above the 123 meter level weigh little more than half a ton, that is wildly incorrect. Blocks on the 196th course, 135 meters above zero level, are 0.58 meter thick, some weighing at least three tons, with corner casing blocks probably reaching four tons. An interior block on the top of the pyramid (upper center of fig. 161) is two courses thick, and must weigh two metric tons. Thus, Houdin's block weights are understated by factors of six to eight. It is hard to imagine how stones this large could be taken through such narrow tunnels.

Houdin says the large corner notches were filled in during the last stage of building by means of both the internal ramp and stones stored externally beside each notch. How could that be done? As a notch was filled in it would cover the internal ramp, would it not? Houdin addresses this problem by saying that "a certain number of support and casing blocks would have to be stored nearby because they could only be put into position at the last moment." He says areas on the faces of the pyramid next to the notches were used as shelves to temporarily store these blocks. But how many blocks would have been stored on these narrow shelves? Let's see.

With the notch dimensions Houdin gives, the volume of each notch is 491 cubic meters. Presuming an average block volume of three quarters of a cubic meter, it would take about 655 two-ton blocks to fill each notch. I estimate that not more than one third of the volume of each notch could have been filled via the internal ramp before access was blocked. The remaining two-thirds, about 437 blocks for each notch, had to be stowed on the steep faces and maneuvered into position from outside the pyramid. Houdin says that these blocks could be placed by sliding them on rollers and finally lifting them into position using "loading tripods." But on what surfaces did the rollers roll? Where were the loading tripods located, what did they look like, and how did they work? We are not told.

As to Houdin's proposal regarding the counterweight system that was employed to hoist the ceiling beams of the King's Chamber, he makes several invalid assumptions:

1. that the builders possessed frictionless bearing at all places where he indicates rollers were used to change direction of ropes. There is no evidence that the builders knew the principle of even Lohner's proposed sleeve bearing, much less frictionless bearings. And if plain bearings such as Hassan's were employed, the counterweight, instead of being 25 metric tons, would have had to be six times heavier.
2. that the roller carriage that transported the counterweight up and down the Grand Gallery would not erode the narrow *limestone* shelves (one cubit wide, but only two-thirds of that width available) on which it ran.
3. that the ceiling of the Grand Gallery, in order to accommodate the counterweight system, was twice as high as it is today. If true, how was the present ceiling created? Houdin doesn't say.

The pyramidion. Houdin imagines it was granite and three meters tall. It would therefore weigh nearly fourteen tons. The method he proposes for raising it is dubious. If it could work at all, the procedure seems unnecessarily complicated. Further, in the last building phase, when it is finally positioned, blocks are inserted *underneath* it to hold it up! Again we have to picture a stone structure being built from the top down.

Lastly, if an internal ramp was used in Khufu's pyramid, why wasn't it used in Khafre's? The upper section of Khafre's pyramid is missing a layer of casing and backing stones about three meters thick, and no signs of an internal ramp can be seen.

The Combination Ramp

Mark Lehner, in *The Complete Pyramids* (1997), proposed a combination ramp, the bottom portion being comprised of two branches. One branch, similar to the Big Straight Ramp, leads from the Great Quarry to a point on the southwest angle of Khufu's pyramid about a third of the way up. The other branch begins at the southeast angle of the pyramid and joins the first branch where the first branch meets the pyramid. From the one-third level, the two branches, now joined as one, spiral up and around the pyramid sides until, near the top . . . he offers no solution.

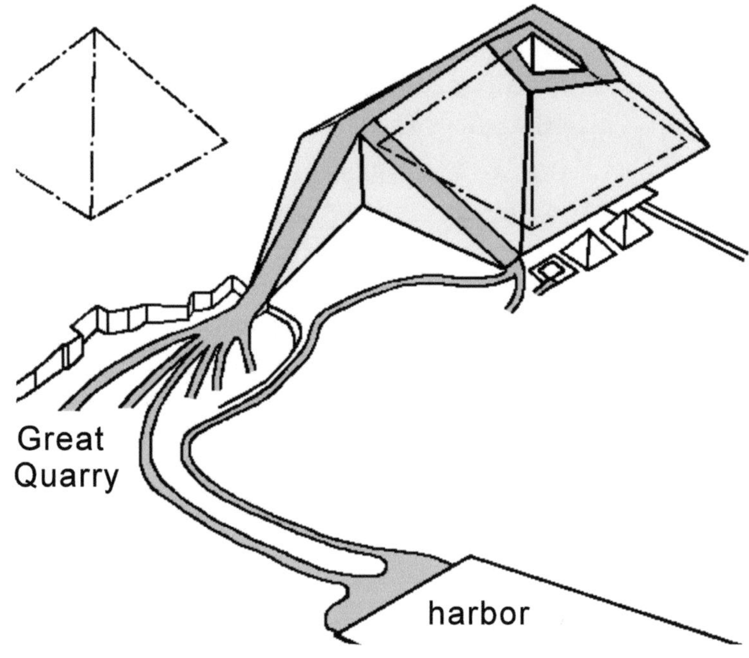

Fig. 247 Lehner's combination ramp. Adapted from Lehner (1997, 204-205).

Lehner's idea has several good points. First, like all straight ramp theories, using a ramp that goes at least a third of the way up Khufu's pyramid is the only way I can see that the ceiling beams of the King's Chamber could have been raised. I don't believe hoisting or levering could do the job. Second, Lehner's spiral ramp is well supported because it extends all the way to the ground. Third, the hauling surface of Lehner's ramp is not as steep as Borchardt-Lauer's big straight ramp, appearing to be about fifteen degrees. Further, the sides of his ramp appear less steep than the big straight ramp, say about 40 degrees. At that angle the ramp sides would not have to be reinforced with masonry to keep them from sliding away.

Lehner deserves credit for being the first, as far as I can determine, to incorporate site topography, quarry, and harbor locations into his proposal.

On the negative side, one problem with foothold embankments and ramps that wrap around the pyramid and extend to the ground is that they require huge amounts of ramp material. It appears that Lehner's ramp volume is greater than the volume of the pyramid. Then there is the problem of workflow. A single spiraling ramp must be a two-way path, one for hauling stones up, and one for return of empty sleds. This means the ramp would have to be wider than its proponents envision. Another problem is that the hip lines become hidden as the pyramid rises, possibly producing for the builders when the ramps are finally removed, a nasty surprise. Last, at the top of the pyramid, like the spiral ramp of Cotsworth, Lehner's ramp would either overlap, become impossibly narrow, or become impractically steep.

With reservations noted, Lehner's idea is at least conceivable.

Another version of the combination ramp was presented by Craig B. Smith in *How the Great Pyramid was Built* (2004). The first part of Smith's ramp is similar to that of Lehner, leading from the Great Quarry to the southwest angle, but much wider. Where it meets the pyramid it begins to spiral upward as does Lehner, but instead of foothold embankments going all the way to the ground, Smith's ramps are small in the manner of Cotsworth.

Smith's idea obviates Lehner's voluminous ramp problem, but not hip line obscuration. Further, in trying to solve what Lehner calls "trouble at the top" he proposed a scenario impossible to achieve, I think, without means for reversing the force of gravity. He recognized that the spiral ramp could not go all the way to the top of the pyramid, and so he makes the ramp stop about 13 meters below the summit.

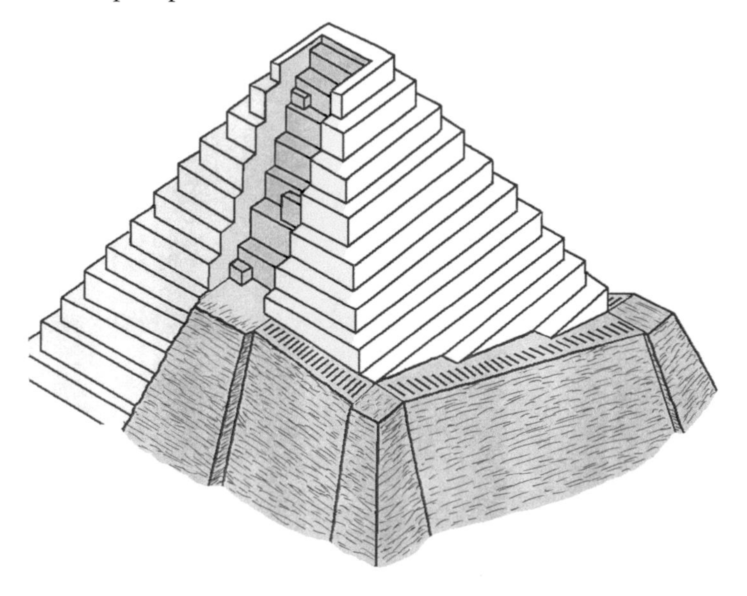

Fig. 248 Smith's spiral ramp and staircase depression at the top of G1. Adapted from Smith (2004, 171).

From that level Smith envisioned the remaining blocks being pushed or pulled up a staircase depressed into the body of the pyramid. He said "once the capstone is maneuvered into place, the staircase is filled in." The remaining blocks "are pushed up from below." It sounds like the capstone is hanging in midair as support stones are placed under it. He then says that the casing stones for the entire pyramid were placed from top to bottom, another procedure I suppose would require the gravity-cancellation device.

My review of existing theories would not be complete if I didn't include a few that range from highly improbable to absurd. Creative for sure, these ideas lack both evidence and logic, but because they seem to be popular in modern media, I thought I should mention a few.

BAD SCIENCE

Pyramid "Concrete"

In *The Pyramids, An Enigma Solved* (1988), French materials scientist Joseph Davidovits and assistant Margie Morris divulged that most of the pyramid blocks are not really quarried stones, but are instead mold-cast "geopolymeric concrete."

I'm not surprised that this revelation gained considerable exposure. It is, to those who have not examined the pyramids, a seemingly plausible explanation for a most perplexing mystery. What *does* surprise me is the length of time Davidovits kept this proposal in the public eye in spite of scientific refutation by eminent geologists like Robert Folk (U of Texas) and James Harrell (U of Toledo – Ohio).

Consider:

1. In observing pyramid blocks I cannot find two identical in size or shape. Thus, if a unique mold was made for each block the main benefit of making a mold, using it for multiple castings, is lost.
2. Why would the masons make their wooden molds such that the cast blocks would appear to have chisel-grooved surfaces? Isn't it more likely that these surfaces were made by chiseling?
3. If each block was cast against the previous block, as Davidovits suggests, how could the builders have put mortar, air voids, and limestone chips between them?
4. Where did the masons obtain 50 million-year-old nummulites and other fossils that they cast into their blocks? These fossils aren't lying around like loose gravel; they became part of the limestone during the Eocene epoch. When confronted with this fact, Davidovits proposed that nummulitic limestone softens when soaked in water, and thereby the fossils could have been obtained for the polymeric mixture. In the interest of objectivity I tested this idea by soaking one of my samples in water for six months. No softening of the stone could I detect.

Said Dietrich and Rosemarie Klemm: "Most probably, the discussion about whether pyramids were constructed using geopolymere will continue. However, it should be clearly emphasized that such theories are nonsense."

Levitation

One of my friends asserts that the ancients must have used mind power (he prefers the term "paranormal" to "psychokinesis") to lift and transport megalithic statues and blocks. When I pointed to some of the carvings and paintings, for example that of Djehutihotep (figure 47), he countered with "maybe a 70-ton problem was reduced to a 7-ton problem." How does one refute such an idea? Let's apply logic. If the Egyptians employed psychokinesis why wouldn't they use it to cancel hauling forces entirely? Why would we find ramps and sleeper-imbedded quarry roads? Why were lever bosses on sarcophagus lids needed? Why would Djoser's crypt-sealing "cork" and Khufu's portcullis slabs need rope grooves? The idea that the Egyptians used levitation seems doubly fallacious, an *argument from ignorance* and an *argument from lack of imagination.*

The absurd

There are many other theories that are not just utterly wrong, but absurdly so. These include the Great Pyramid being raised by floating blocks up its side on barges via a series of locks, lifting the blocks with huge kites, and floating blocks to the pyramid on barges on a surrounding artificial lake, the water level of which is gradually raised by increasing the height of a dam.

Detouring from stone-raising for a moment, but staying with the paranormal, in the 70s we suffered a craze of "pyramid power." The energy-focusing pyramid shape could keep razor blades sharp, preserve food indefinitely, increase rabbit production, and improve meditation. A friend informed me that the needle of a magnetic compass would spin in the King's Chamber of G1. And so, as this was easy enough to test, I packed a compass for my 1978 sojourn. The compass worked normally in the King's Chamber and on top of the pyramid.

A friend said he was able to bend a normally-unbendable spoon; another said she was able to bend a large nail. When I asked for demonstrations of those abilities they said they could perform those feats only "under the right conditions." "What were those conditions," I asked? Well, they had to be in the right place (e.g. an institutional retreat where "vortex energy" was focused), in the right state of mental preparedness, and under the tutelage of an expert. I asked if these tests of mental adroitness were scientifically controlled. "Say what?" May I be pardoned for remaining skeptical?

RAMPS-PLUS-HOISTING

Now that I have excoriated others' theories it's time to give you a shot at me.

The stone raising method that is most in harmony with history, evidence, and known technology combines evolving ramps with a final hoisting phase. After years of sketching ideas and making clay and computer models with ramps of every sort I settled on that shown in figure 249. Mind you, I do not call this my most favored idea; I call it my least objectionable idea.

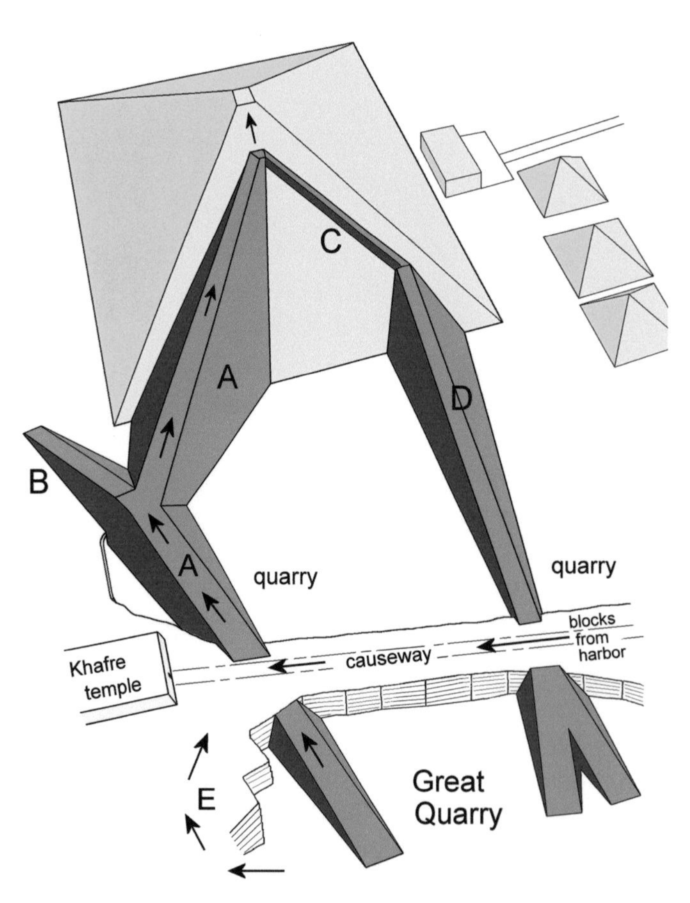

Fig. 249 The ramp arrangement at near-completion of Khufu's pyramid.

The drawing shows the ramp configuration when Khufu's pyramid is almost complete. Ramp **A** is the main ramp over which stones are hauled. Ramp **B** is a run-off for the workers, especially those who had previously hauled up the megaliths for the King's Chamber, Queen's Chamber, and Grand Gallery. Ramps **C** and **D** are used for returning empty sleds to the quarry. Ramp **D** was used in an earlier phase as another path to raise blocks. Hauling path **E** may have been needed as blocks were taken from deeper parts of the Great Quarry. Tura casing blocks are being dragged up from the harbor along the causeway that leads to Khafre's mortuary temple.

It is likely that casing blocks were sorted, roughly finished, and temporarily accumulated in a staging area near the harbor or close beside the path or paths to the pyramid.

I'll get to building phases in a moment.

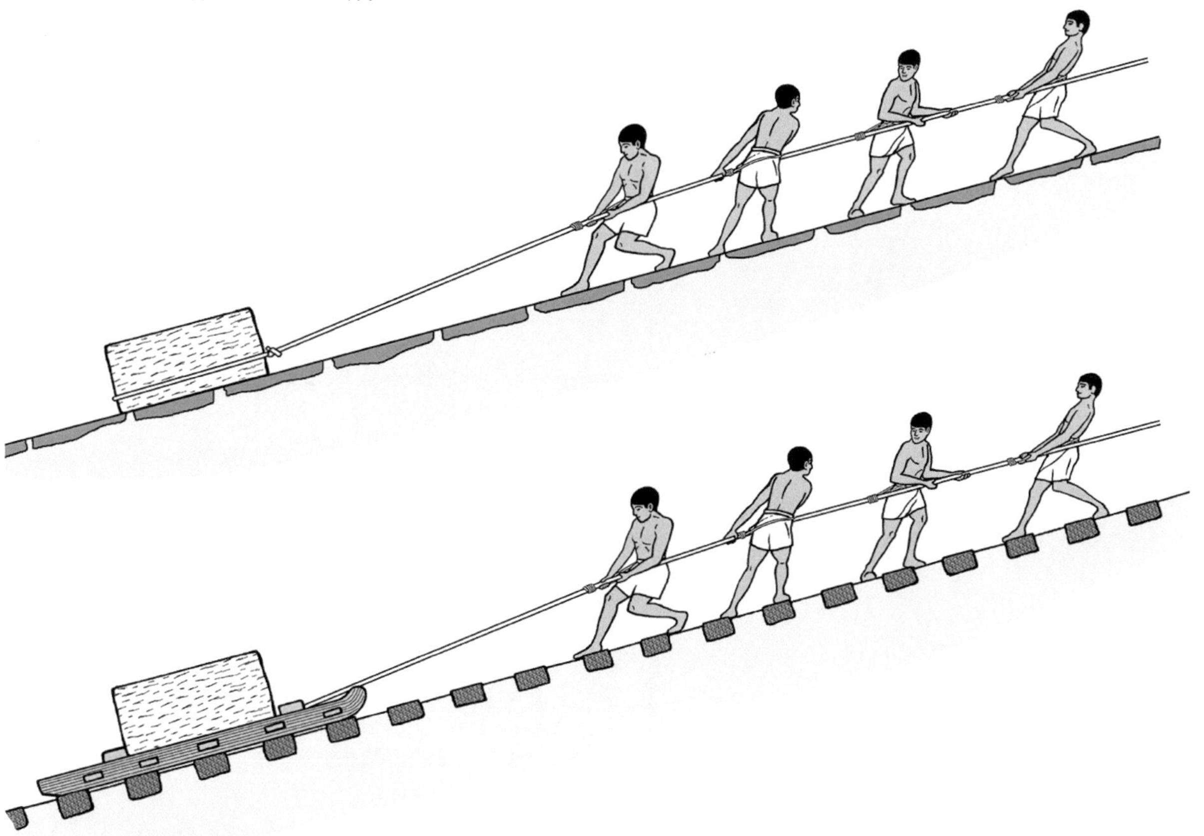

Fig. 250 Two methods by which blocks could have been dragged up the main ramp.

Perhaps blocks were dragged directly over stone paving slabs. No need for sleepers or sleds. This would have worked well at Giza. The surface between quarries and pyramids is rock. Thus, there would have been no concern for sled runners sinking into sand or soft ground. Paving slabs would have been needed only for ramp surfaces, and since these slabs are man-portable they could have been easily moved when ramps were reconfigured as the pyramid rose. The process would have been simpler if sleds were not needed. Less weight has to be hauled up the ramps. Only ropes have to be carried back to the quarry. The problem of matching sled size to block size vanishes.

Figure 251 shows the pyramid at 12.5 percent (one-eighth) of its final height, a little over eighteen meters. At this level, one third of its final volume has been established.

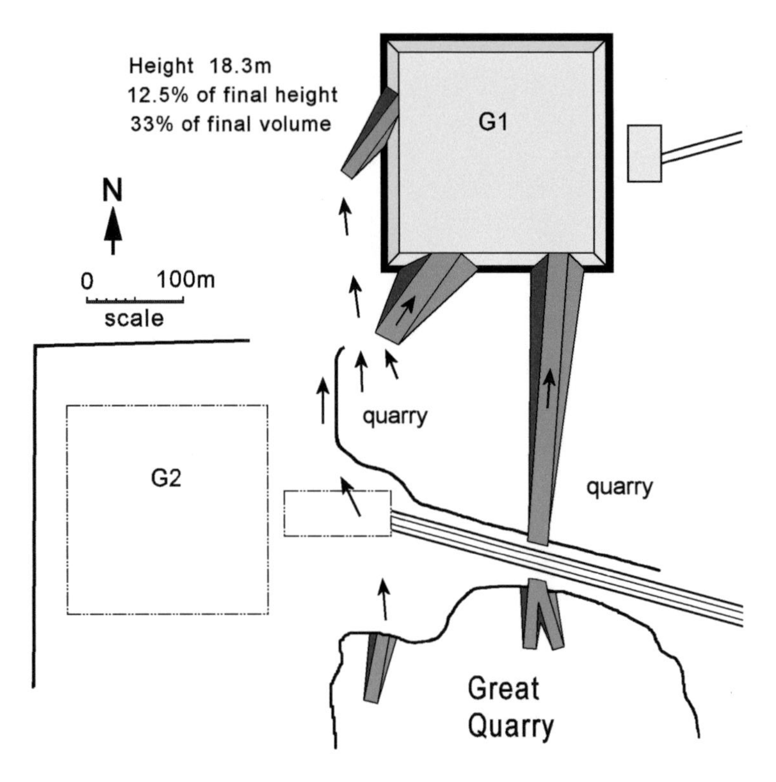

Fig. 251 Ramp layout at one-eighth of final pyramid height.

Two ramps are disposed against the south face of the pyramid, another on the west face. Many other ramp configurations are possible, even likely, at lower levels of the pyramid. Three ramps (or more) allowed multiple paths for the stone-haulers, both to the pyramid and for return to the quarries. At this height, blocks had to be placed at a rate of about one every two minutes. Many stones have been taken from the quarry immediately south of the pyramid. Others are being taken from the Great Quarry. The fact that the G2 causeway was not cut through by ramps from the Great Quarry to G1 suggests that Khafre's pyramid was already planned. Casing stones coming from the harbor arrive at the main ramp via the Khafre causeway. Work has begun on Khufu's mortuary temple and the three small pyramids on the east side of G1.

Figure 252 shows the ramp layout when the pyramid had reached seven-sixteenths of its final height, or 64.1 meters. This height corresponds to the top of the uppermost relieving space above the King's Chamber. The main ramp has reached an angle with the horizon of twelve degrees.

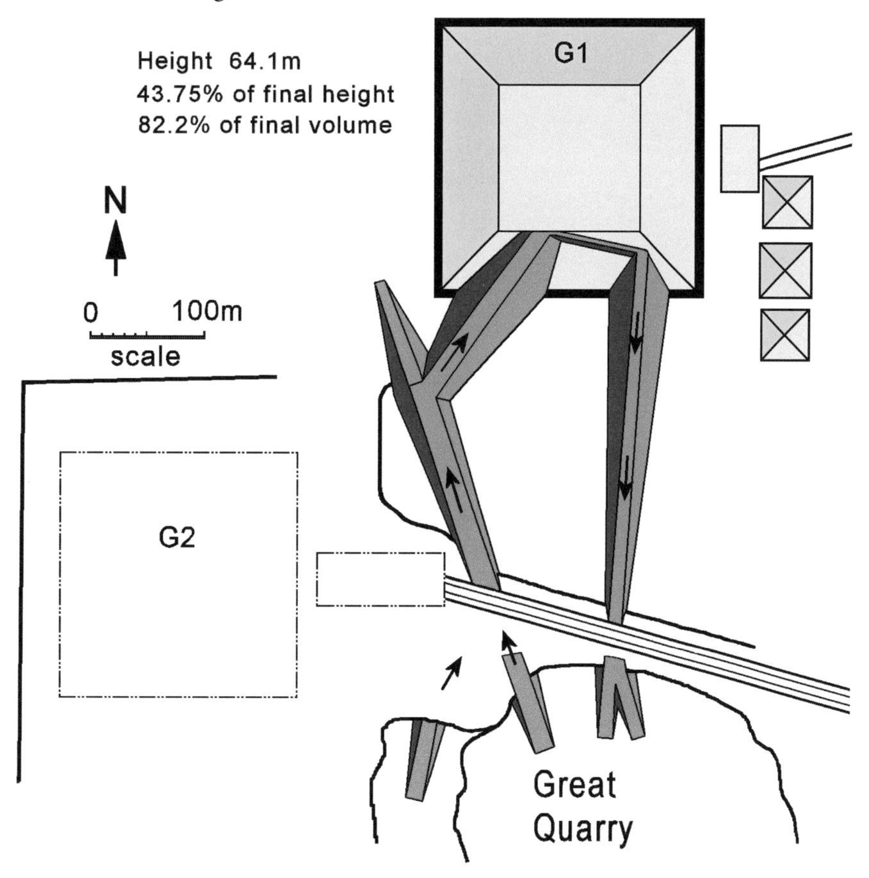

Fig. 252 Ramp layout at 7/16ths of final pyramid height.

Stone is being taken exclusively from the Great Quarry. Eighty-two percent of the volume of the pyramid has been completed. Only the main ramp is being used to raise stones. A small mud-brick ramp along the face of the pyramid is used by workers returning to the quarry by way of ramp **D** in figure 249. Khufu's mortuary temple is complete, as is pyramid G1a. Work is underway on pyramids G1b and G1c.

The granite rafters of the King's Chamber are the heaviest stones, as far as we know, that are high in the pyramid. One of the beams in the lowest ceiling, 49 meters above zero level, is 8.5m long, 2.1m high, and 1.5m wide (Maragioglio & Rinaldi). Its mass is nearly 68,000 kg, or 75 U.S. tons. These blocks were mounted on sleds and brought from the harbor to the base of the ramp via what was, or would become, the causeway of Khafre's pyramid. To raise these enormous blocks the surface of the ramp was coated with Nile mud. Water poured ahead of the sled runners reduced the coefficient of friction between sled and ramp to about 0.15. Even so, the heaviest block would have required, I figure, at least 730 workmen.

Figure 253 shows the ramp when the pyramid had reached 78.5 percent of its final height, or 115 meters. The upper part of the main ramp (from where it turns 45 degrees to the right) has reached a slope of fifteen degrees. At the 115 m level, only one percent of the pyramid's final volume remains to be established. For the last 32 meters the stones are hoisted along the pyramid face.

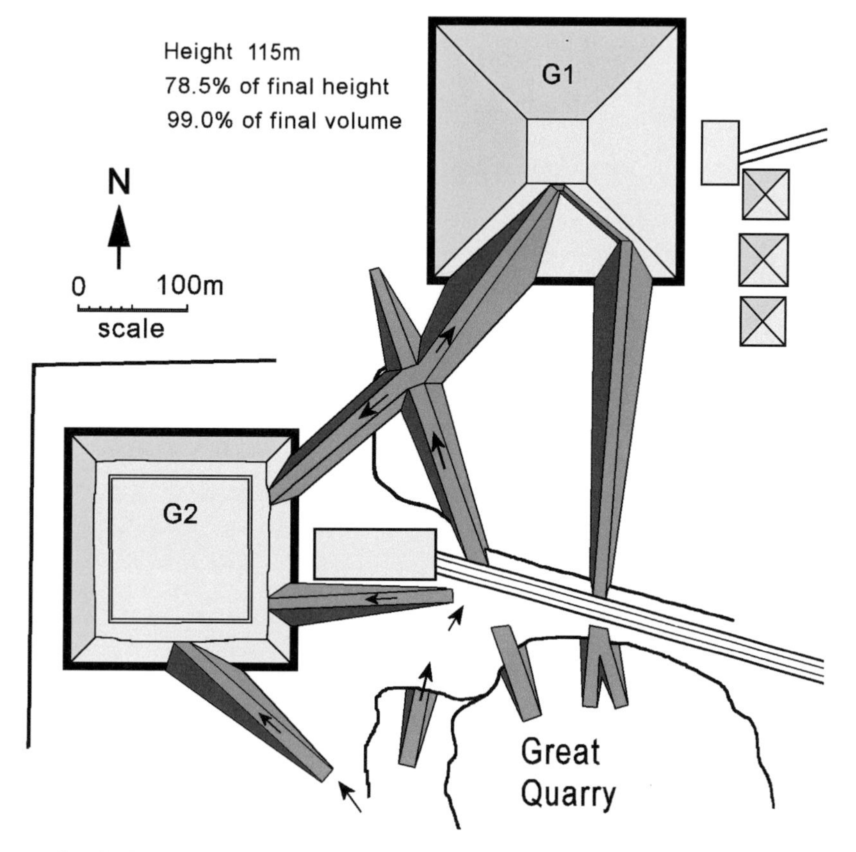

Fig. 253 Ramp layout at 78.5 percent of final pyramid height. Only one percent of its final volume remains to be established.

I now come to what Mark Lehner aptly called (1997, 222) "trouble at the top," raising and placing the blocks at the top of the pyramid where there was little room for workmen to stand. Casing stones in the form of stair-steps, i.e., not planed to the final pyramid slope, would have been especially useful here. Figure 254 shows how workmen may have hoisted the pyramidion while arrayed on the opposite face of the pyramid.

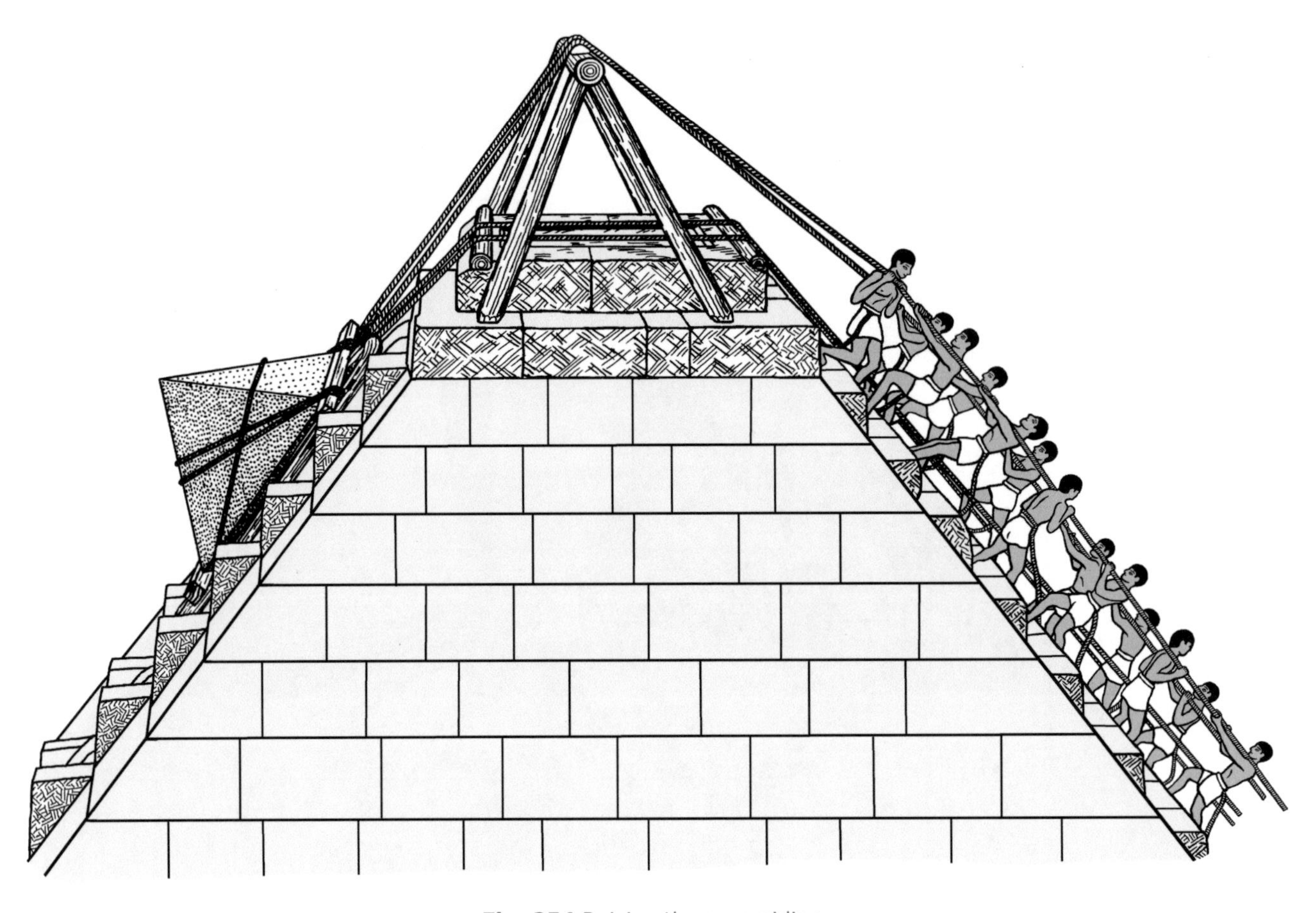

Fig. 254 Raising the pyramidion.

The hoisting ropes go around Hassan's stone bearings at the uppermost cross-beam. The ropes lower down go over half-logs, without bearings, and are used to hold the load against slippage when the hoisters rest or climb back up for another hauling cycle. All workmen are attached to the pyramid with safety ropes that loop around the top two courses of the pyramid, ensuring that if a rope breaks they would not tumble to their deaths.

In surmising the stone-raising method for G1 I thought I should also consider G2. Thus I will add one more piece of evidence that may or may not be a clue. On the northern border of G2, at its northwest corner, there is a row of mortar-filled holes in the pavement that leads away from the corner. The holes make an angle of about ten degrees with the base. See my photograph, figure 255. For orientation, the small arrow at the upper left corner points to the roof of the boat museum at the south base of G1.

Fig. 255 The pavement holes at the northwest angle of G2.

This row of holes is independent of the holes I mentioned in chapter nine, the holes Lehner believes are layout holes for the pyramid base. The diagonal row of holes are the same size, with the same spacing as the layout holes, but do not seem correlated. What purpose did they serve? Perhaps they held poles which buttressed a timber-walled ramp. Just a guess. I show these holes, with diameters greatly exaggerated, in figure 256.

In figure 256 I show the ramp arrangement for G2 as it reached the same phase of construction as did G1, i.e. when the hoisting operation began.

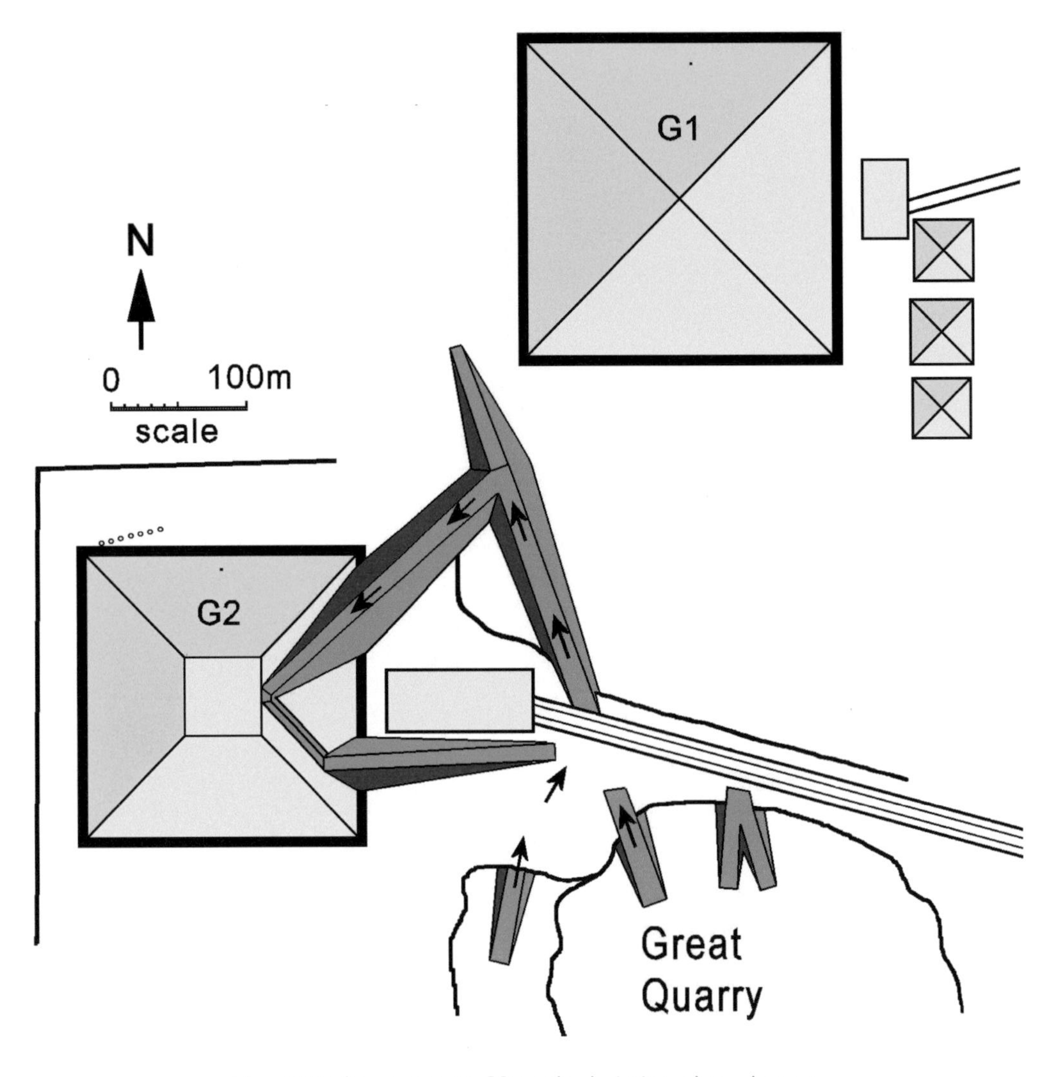

Fig. 256 The ramps on G2 as the hoisting phase began.

The upper portion of the main ramp for G1 has been removed and reused for the upper portion of the main ramp for G2. Thus about a third of the volume of G2's main ramp did not have to be built for G2. G2's main ramp is somewhat longer than that of G1 because G2 sits on ground about ten meters higher than G1.

The ramps I have described have large slopes. Intuitively you might think that these are excessively steep. Shouldn't we be looking for a ramp with the lowest possible slope?

Yes, if we are trying to minimize the dragging force required, and thus the number of draggers per block. But no if we are trying to minimize the amount of *energy* (or *work*) required to raise a block a given height. The reason has to do with energy lost to friction. Incidentally, I will use the terms *energy* and *work* interchangeably because for the subject at hand they are the same thing.

Referring to figure 257, the force required to haul a sled-borne stone of weight **W** up a ramp is the sum of the force component L (equal to W sin θ) required to overcome gravity and the force required to overcome friction of the sled runners (μN) as they slide over sleepers embedded in the ramp material.

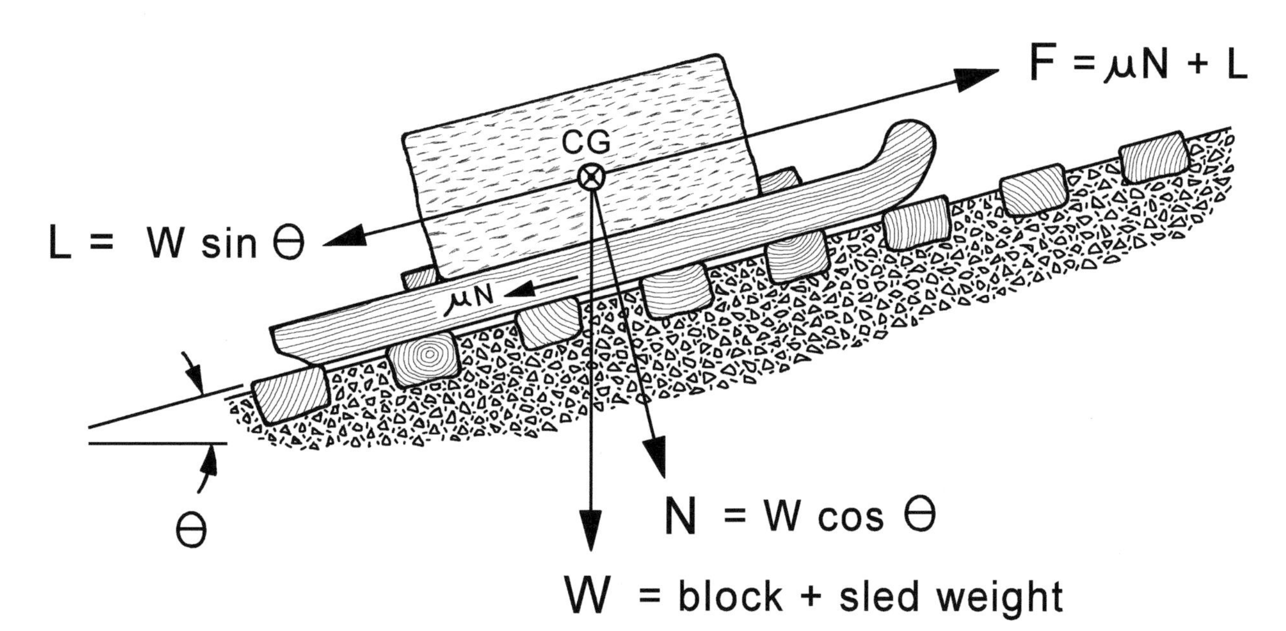

Fig.257 Forces on a block-laden sled.

Angle theta **θ** is the ramp angle. The symbol **μ** is the coefficient of friction between the sled runners and sleepers. **N** is the normal force, the component of the weight that is directed perpendicular to the surface on which the load is sliding, and is equal to W cos **θ**. The lower the ramp angle the greater is the percentage of frictional drag to the overall dragging force required. Energy lost to friction is work completely wasted.

For example, using 0.30 as the coefficient of sliding friction of an oak plank sliding cross-grain on another oak plank, the force required to drag a three-ton stone and its 300 pound sled up a five degree incline is 2,432 pounds-force (10,817 N). If the workers could pull with 70 pounds-force each (311 N), thirty-five workers would be needed. The force required to drag the same load up a fifteen degree ramp is 3,456 pounds-force (15,372 N), 42 percent more force, requiring forty-nine workers. However, the energy expenditure in hauling the sled-borne stone plus the workers' own bodies up the steeper ramp to the same vertical height is only 60 percent of that required for the five degree ramp, so from an energy efficiency standpoint the steeper ramp is the better choice.

The main problem with steeper ramps is assuring that the block draggers can gain traction with their feet on the ramp. Any slippage would waste energy. Perhaps the workmen used protruding sleepers for foothold purchase.

I addressed pyramid building times in the section on Khufu's pyramid in chapter seven. There I estimated that Khufu's pyramid was probably erected over a span of about thirty-four years. Recall that Hawass stated (2011) that he believes Khufu reigned 30-32 years. I also proposed that Djedefre spent most of his eight-year reign completing work on Khufu's pyramid. Thus, for the following analysis of stone raising methods I will use thirty-four years for the erection of Khufu's pyramid.

GUESSTIMATING

To keep things simple I am going to compare only two stone-raising methods, lever-jacking and ramp-plus-hoisting. These I think are the two best ideas. Also for simplicity I will base my figuring on building the largest pyramid, Khufu's.

Could the Great Pyramid have been raised in thirty-four years by either of these methods? If so, how many workers would have been required? To answer these questions we must first estimate the size and number of blocks that had to be raised. Then we would know the minimum work (energy) that was needed no matter what stone-raising method was used. Finally I will look at the efficiency of both stone-raising methods to see if either could have accomplished the job in thirty-four years (or fewer).

If you don't have the desire or patience to consider the following figures and estimates (requiring, I promise, only arithmetic) you may skip to the section below titled **Summary of human and time requirements for erecting G1**. With that caveat, let's proceed.

It has long been said that the Great Pyramid contains 2.5 million blocks averaging 2.5 tons each. But these numbers assume that the pyramid is a homogenous structure of tightly-packed limestone blocks.

That is not the case.

Let's start with the total volume of G1, the volume within the original outer surface from zero level upward. A pyramid with 440 cubit base length and 280 cubit height has a volume of 2,593,058 cubic meters (91,571,250 cubic feet).

To get the total volume of blocks to be raised we must subtract the volume of the core massif, part of the Giza plateau that rises within the base of the pyramid. It is six meters above zero level at the top of the Grotto, and probably another two meters higher at the center of the pyramid. Thus it comprises perhaps eight percent of the volume of the pyramid. We must also subtract the internal apartments, gaps, sand, clay, mortar, stone chips, and small limestone chunks, features we can readily observe.

There's more.

In 1986, French architects Gilles Dormion and Jean-Patrice Goidin, declaring they had detected voids west of the corridor leading to the Queen's Chamber, got permission from the Antiquities Service to drill three holes in that direction. After nine feet the drills found only pockets of sand.

In 2015, Japanese and French scientists, using muon-tomography scanners, reported a 30-meter-long "void" above the Grand Gallery. Muons are cosmic rays that are scattered (deflected) more when passing through masses of greater density, so they can reveal density differences. The scientists interpreted their scans as indicating that the lower-density volume is an *air* pocket. But could it be a *sand* pocket? Quartz sand (SiO_2) is not only less dense than limestone ($CaCO_3$), it is a different chemical compound. Sand may have been used as both a stress-distributing agent above the largest room in any pyramid, and a time-saving expedient (there and elsewhere). Sand pockets in non-critical areas, protected from erosion by rainwater, would have saved much labor.

Sand pockets and the other non-block components I listed above probably total twenty-two percent of the pyramid volume. All told, I figure quarried limestone blocks constitute, at most, seventy percent of the volume of G1. Even so, the pyramid would contain nearly two million blocks averaging 2.35 metric tons (2.6 U.S. tons). This is the *maximum* number of blocks in G1 for which we have to find a reasonable method by which they could have been raised.

The total can be further apportioned into blocks small enough to be carried by a 12-man team, and larger blocks which could not be carried. The number of blocks too heavy to be carried becomes one-and-a-half million. Those which could have been carried (less than 555 kg) number a half-million.

If G1 was truly raised in thirty-four years we can say something about the rate at which blocks were raised and placed. Hawass says (2006, 175) "the workers took one day off in every ten," worked ten-hour days and had "numerous feast days" (holidays). We'll say the stone raisers labored ten hours a day for three hundred days each year (Incidentally, the work day may not have included mid-day hours when the sun was most oppressive). Ten hours times 300 days times 34 years yields 102,000 hours of stone-raising time. If we divide this time into the number of stones to be raised (2,000,000) we find that, on average, about twenty blocks had to be raised (and placed into position) each working hour, about one block every three minutes. This rate is misleading, however, because block raising work (and time) increases with height raised. I figure the placement rate at the lowest courses had to be more like one block every two minutes, diminishing to a block every four to five minutes at the top. Constructing internal apartments and laying casing stones would take more time than rough core work, but that extra time would be offset by the faster rate at which man-carried stones were raised and placed.

Now let's see how much work, at minimum, is required in raising two million stones.

The minimum work in erecting G1 is represented by its *gravitational potential energy* (GPE). Potential energy is the ability to do work, usually pertinent when an object has the potential to "fall" in a gravitational field. In our case, GPE is the energy given to an object that is raised above the ground. A simpler way to look at GPE, in an object near the earth's surface, is that its GPE is the product of its weight times the distance of its *centroid* (center of mass) above the ground. The centroid of a pyramid lies at one fourth of its height, so for G1 its centroid is 481/4 = 120.25 feet above zero level. If G1 was a homogenous solid with a density of 144 pounds per cubic foot and a volume of 91,571,250 cubic feet, its weight would be 13,186,260,000 pounds, and its GPE about 1.59 trillion foot-pounds. If we subtract the core massif and voids, etc., the GPE becomes roughly 1.31 trillion foot-pounds (1.78 trillion joules). That figure represents the minimum amount of work needed to raise G1 if the stone-raising method was 100 percent efficient, which, subject to the second law of thermodynamics, cannot be.

Now to *efficiency*, how much work is accomplished compared to the effort expended, output divided by input. Since we are looking for the actual work needed (work input) to raise G1, we will rewrite our equation as:

Work Input = Minimum Work Output / Efficiency

I reckoned the minimum work output was 1.31 trillion foot-pounds.

If I can estimate the efficiencies of the two stone-raising methods I can use the above equation to obtain the work input needed for the job. I can then display my result either as the number of workers required to raise the pyramid in thirty-four years or the number of years required to build the pyramid for a given number of block-raisers.

The evolving-ramp-plus-hoisting method entails two phases (ramp and hoisting), but because the ramp takes care of 99 percent of the pyramid volume, I'll compare only ramp-raising to lever-jacking. The slope of the ramp is small at the start of the building, and increases with increasing height. To simplify matters I'll consider just two data points for the ramp, when its angle reached five degrees and fifteen degrees. I will spare you the calculations and just report my results.

Stones can be raised with the five degree ramp at about five vertical feet every minute. The rate becomes about eight vertical feet per minute with the fifteen degree ramp. As noted earlier, Hodges found the block raising speed for lever-jacking is 0.20 feet per minute, a rate 25-40 times slower than dragging stones up a ramp.

Lever-jacking might reduce the speed advantage of ramp-hauling because it can employ more tracks for raising the stones, but the advantage of more tracks is offset by the need to build stone staircases over which the blocks are lever-jacked. When the stone staircases are being enlarged, the stone-jacking rate of *pyramid* blocks is reduced. Ramp expansion, conversely, can proceed almost simultaneously with the hauling of blocks because ramp material can be hauled, for example, by women and children, and at night. With consideration of number of tracks available (a positive) and auxiliary work required (a negative), lever-jacking is, *at minimum*, ten times slower than ramp-hauling.

Let's look at the *power* requirement for ramp-raising the stones. I figure that the output required per workman in dragging a block up a fifteen degree ramp is around 3,265 ft-lbs per minute, or 74 watts. Studies have shown that a well-conditioned man can generate this power for eight hours a day if using his leg muscles (climbing stairs, towing a barge, bicycling) and somewhat less for the same period if using his arms (turning a winch, pulling a lever). An output of 3,265 ft-lbs per minute for eight hours would be 1,567,200 foot-pounds of work. Here I assume that eight hours of work would be performed in a ten-hour work day.

Humans are fueled by food calories, each of which contains 3,088 foot-pounds of energy. But because human efficiency using leg muscles is only 20 percent, a man must burn five calories to produce one calorie of output. So, to produce 1,567,200 foot-pounds of work per day a man must burn five times 1,567,200 divided by 3,088, or 2,538 calories. This must be added to the 2,250 calories a 150 pound man must consume to maintain his body weight (stay alive and not lose weight). The total minimum daily caloric intake for such a workman is thus 4,788 calories, a value not unusually high for heavy labor (e.g. farm work) today. Ergo, the sled haulers had to eat well.

Using a lever is more mechanically efficient than dragging an object over the ground because with a lever less energy is lost to friction. However, because arm muscle efficiency is less than leg muscle efficiency, the two methods become approximately equal in overall efficiency. Efficiency, then, for both stone-raising methods is about twenty percent.

The total **work input** then required of both stone-raising methods becomes 1.31 trillion foot-pounds divided by 0.20, which equals **6.55 trillion foot-pounds**.

If we now divide the total work input (6.55 trillion foot-pounds) by the 1,567,200 foot-pounds of work per day that each ramp-hauler could provide, we find that 4,180,000 man-days were needed to raise G1.

Next we have to make a guess at how many haulers were employed to haul the stones.

As a first approximation, let's say that 1500 workmen were used for ramp-hauling the stones. At any given moment there are 1000 men in twenty 50-man crews hauling stones to the pyramid, up the ramp and across the top to where the stones were delivered. Half as many haulers are returning empty sleds to the quarry because they can walk twice as fast downhill dragging only sleds.

With 1000 men hauling stones at a constant rate from the quarry to where the stones were delivered, 4,180,000 man-days divided by 1000 men yields 4,180 working days for raising the stones of G1. If the workers worked 300 days each year, then 4,180 divided by 300 equals roughly fourteen years required for building the Great Pyramid.

Fourteen years would be the most labor-intensive, shortest-time scenario for ramp hauling. This is about the time estimated by Egyptologist John Romer, who said (2007, 118), "In less than fourteen years, more than 5 million tons of cut stone blocks were hauled up out of that quarry [the Great Quarry] and set up on the Great Pyramid: a theoretical total of around 2,080,000 standard 2.5 ton blocks." Romer's figures equate to a "standard block" (if there *were* such a thing) being placed at a rate of nearly one-per-minute over the fourteen year period. Romer continues: "… it appears that workloads were very much higher at the beginning of the work than at its ending. It follows, then, that working the notional ten-hour day for 300 days each year, in those first few years during which half the Pyramid was built, more than 12,000 souls would have been required to labour in the Giza quarry, a colossal number in itself, quite apart from those working on the Pyramid."

But are these estimates realistic? I think Romer overestimates the number of quarrymen and pyramid workers required, and he and I both underestimate the minimum building time.

Consider a more conservative scenario.

Let's say the haulers produced only fifty percent of the power I estimated above. If 500 permanent haulers worked at this pace for 300 days every year, and were augmented by 1000 seasonal haulers working at the same pace for 120 days (4 months) of every year it would have taken 33 years to erect the pyramid.

An infinite number of combinations are possible, but it appears reasonable that G1 could have been raised by ramp-hauling in not more than thirty-three years, perhaps a few years less.

With lever-jacking, we would still need hauling crews to get blocks from the Great Quarry to the bases of the four stone staircases. One of these is on the north side of the pyramid *opposite* the quarry. Another would prevent the building of the mortuary temple on the east side of the pyramid, and the third is preventing the building of the mastaba tombs on the west side. Further, the rate at which lever-jacked stones are raised cannot be increased by asking workmen to work faster or assigning more workmen to the job, so the many variables that can be manipulated in ramp-raising are not as broad in lever-jacking. Because lever-jacking is ten times slower than ramp-hauling it would take, under the same conservative rules as with ramp-hauling, 300 years to complete G1 and its four stone staircases (not to mention the time it would take to dismantle the staircases).

Now for an estimate of stone-raising employment. I'll use the conservative scenarios above. For ramp-hauling I figure we must add about 350 *ramp builders* to our total. The maximum level of employment needed for ramp-raising the pyramid blocks is then 1,500 plus 350, or 1,850 workers. The total for lever jacking, considering auxiliary work of stone staircase construction, cribbing, plus workmen dragging blocks from the quarry to the bases of the stone staircases, I estimate would be about twenty percent more, or 2,220.

So we have with ramp-hauling about 1,850 workmen raising the stones, versus 2,220 with lever-jacking. In both cases, another 600 workmen would be freeing blocks in the quarry and loading them onto sleds. Another 300 would be unloading, fitting, and placing the blocks, including chunk and mortar-filling. About 1,000 workmen would be needed to quarry and transport stones from Tura and Aswan. Both methods employ roughly the same number of workers, but lever-jacking still takes ten times longer.

One last point favoring ramps for building most of the pyramid: Ramps would have provided more than just paths for stone haulers. The block fitters on top of the pyramid would have needed a lot of drinking water during the working day, as anyone who has hiked the Grand Canyon in July can attest. Many other tasks required movement of people to and from the upper reaches of the pyramid. The transit of these folks would have been much easier by means of a ramp than by climbing up and down stairways.

Summary of human and time requirements for erecting G1

Ramp-hauling would require about 3,750 quarrymen, ramp-builders, sailors, and stone-raisers to complete the job in 33 years. I'm going to increase this number to 5,000 to account for contingencies I haven't considered. Lever-jacking, under the same rules, would require at least as many workmen and 300 years, so it is not a reasonable alternative.

To the total of 5,000 workmen we must add at least 15,000 support people (including women and children). These would include carpenters making and repairing sleds, rope makers, food preparers, water carriers, gypsum miners, mortar-mixers, ship-builders, dock workers, boat crews, supervisors, medics, scribes, and jobs I haven't imagined. My total of twenty thousand people working on Khufu's pyramid is within modern estimates. In an article in the November 2001 issue of *National Geographic* titled "The Pyramid Builders," Virginia Morell says "Hawass and Lehner estimate that the feat – quarrying, transporting, and fashioning the seven million cubic yards of stone for the three pyramids and adjoining structures – was accomplished with a workforce of only 20,000 to 30,000 men." If it turns out that there are purpose-designed pockets of sand within the pyramids, especially G1 and G2, then my estimates for blocks required, total building time, and manpower are too high by as much as ten percent.

Weaknesses of my idea

The hoisting phase, though required only for the last one percent of the pyramid volume, seems dangerous, especially if hoisters had to hang off the opposite face of the pyramid while raising the last few courses.

Without foothold embankments there would not be much maneuvering space for the casing placers. In rolling their blocks into position there would be little margin for error – if a block protruded too far beyond the pyramid face it would be difficult to pull it back into proper position.

The variability in block size would demand perhaps six sled sizes so that each sled was in the right proportion to its load, not so large that unnecessary weight was being lifted. Constantly matching sled size to block size is complicated, even if, at times, two or three small blocks were carried on the largest sled. This problem goes away if sleds were not used.

Though I have argued that nucleus structures were probably not advanced significantly above the casing work, except at Meidum, I could be wrong. If a nucleus structure was advanced substantially above the casing/filling work then at least part of the ramp for building the nucleus would have had to be removed so casing/filling could be placed where the ramp had been. That would have entailed a lot of extra work.

Last, like all other stone-raising theories, my musings are tenuous, certainly wrong in detail. Unless we find drawings or descriptions by the builders, the stone raising method (or methods) will remain a mystery. Nevertheless, because my fallback purpose was to find a *feasible* method by which the stones could have been raised, I believe I have accomplished that objective.

11

EPILOGUE

With regard to my overall mission, a friend asked: "What is at stake?" I replied that I wanted to dispel the popular notions that the pyramids were built by slaves and that pyramid building was a gigantic waste of time and resources. I also wanted to show why I admire the engineering savvy, creativity, and persistence of the builders.

Was pyramid-building a waste of time and resources? Who are we to judge? Folks may look back some day and question *our* activities: world wars, species extinction, and environment degrading. The organization required for pyramid construction was certainly a beneficial precursor, planned or not, in building the state of Egypt.

I hope you will forgive my lack of conviction on how the builders raised the stones. A challenging journey is often more fulfilling than a finish-line crossed. I've had fun. And maybe I've provided a few clues that will let *you* solve the mystery.

Perhaps it is not so intractable after all. Long ago, while putting pencil to paper on my drawing board (pre-computer), a fellow designer many years my senior (he must have been in his fifties) turned to me and said "Robert, I think I know how they built the pyramids."

"Yeah? How?"

"They dug a big hole in the ground in the shape of an inverted pyramid, filled it with blocks, and flipped it over."

"Great!"

ACKNOWLEDGEMENTS

I thank my mentor Professor Hans Goedicke of Johns Hopkins University for his enthusiastic support of my investigation of the pyramids. Discussions with Dr. Goedicke and his wife Lucy provided the early kindling I needed to build the fire. I am grateful to my friends at Black & Decker whose encouragement over the years kept dilatory me working on the manuscript. Thanks to Elie Rogers and Pete Petrone of *National Geographic* for soliciting my help on the Khufu Boat project, and introducing me to my good friend Pieter Tans of NOAA. I thank my daughter Sheri for her photography, her husband Scot for copy editing the manuscript, and Annie Woodall and Clare Munroe for their beta reading.

BIBLIOGRAPHY

Arnold, D. *Building in Egypt*. New York & Oxford: Oxford University Press, 1991.
Badawy, A. *A History of Egyptian Architecture*. Berkeley & Los Angeles: University of California Press, 1966.
Ballard, R. *The Solution of the Pyramid Problem*. New York: John Wiley & Sons, 1882.
Baumeister, T. and L. Marks. *Standard Handbook for Mechanical Engineers*. New York: McGraw-Hill, 1967.
Bauval, R. and A. Gilbert. *The Orion Mystery*. New York: Crown Trade Paperbacks, 1994.
Brier, B. and J-P Houdin. *The Secret of the Great Pyramid*. New York: Harper, 2008.
Brunton, G. *Lahun I*. London: British School of Archaeology in Egypt, 1920.
Bureau of Naval Personnel. *Principles of Naval Engineering*. U.S. Navy, 1966.
Clarke, S. and R. Engelbach. *Ancient Egyptian Masonry*. London: Oxford University Press, 1930.
Cole, J.H. *Determination of the Exact Size and Orientation of the Great Pyramid of Giza*. Cairo: Government Press, 1925.
De Morgan, J. *Fouilles a Dahchour*. Vienna: Adolphe Holzhausen, 1895.
Edgar, C.C. *Sculptors' Studies and Unfinished Works*. Cairo: Service des Antiquites de l'Egypt, 1906.
Edwards, I.E.S. *The Pyramids of Egypt*. New York: The Viking Press, 1972.
El-Baz, F. "Finding a Pharaoh's Funeral Bark." *National Geographic*, vol. 173, no. 4, April 1988.
Emery, W. *Archaic Egypt*. Harmondsworth: Penguin Books, 1961.
Engelbach, R. *The Aswan Obelisk*. Cairo: Service des Antiquites de l'Egypt, 1922.
Fakhry, A. *The Monuments of Sneferu at Dahshur - The Bent Pyramid*. Cairo: Antiquities Dept. of Egypt, 1959.
———. *The Pyramids*. Chicago & London: The University of Chicago Press, 1961.
Gantenbrink, R. *The Upuaut Project*. http://www.cheops.org/ 1999.
Gillings, R.J. *Mathematics in the Time of the Pharaohs*. Cambridge, Mass: M.I.T. Press, 1972.

Goedicke, H. "Some Remarks on Stone Cutting in the Egyptian Middle Kingdom." *JARCE* 3. 1964.
Goneim, M.Z. *The Buried Pyramid*. London: Longmans, Green and Co., 1956.
———. *Horus Sekhemkhet*. Cairo: Service des Antiquites de l'Egypt, 1957.
Goyon, G. *Le Secret Des Batisseurs Des Grandes Pyramides Kheops*. Paris: Editions Pygmaloin, 1977.
Greaves, J. *Pyramidographia*. (new edition) Baltimore: The Maryland Institute Press.
Hamilton, W.R, A.R. Wooley, and A.C. Bishop. *A Guide to Minerals, Rocks and Fossils*. New York: Crescent Books, 1974.
Hancock, G. & R. Bauval. *The Message of the Sphinx*. New York: Crown Publishers, Inc., 1996.
Hassan, S. *Excavations at Giza*, vols. VI - X. Cairo: Government Press, 1960.
Hawass, Zahi and M. Lehner. "Builders of the Pyramids." *Archaeology*, volume 50, no. 1, 1997.
Hawass, Zahi. "Tombs of the Pyramid Builders." *Archaeology*, volume 50, no. 1, 1997.
———. *Mountains of the Pharaohs*. New York: Doubleday, 2006.
———. *Secrets from the Sand*, New York: Harry N. Abrams, Inc., 2011.
Hayes, W. *The Scepter of Egypt*. New York. Harper & Bros, 1953.
Herodotus. *The History of Herodotus*. (G. Rawlinson Transl.). New York: Tudor Publish. Co., 1932.
Hodges, P. *How the Pyramids Were Built*. Longmead: Element Books, Ltd., 1989.
Isler, M. "Ancient Egyptian Methods of Raising Weights." *JARCE* 8, 1976.
———. *Sticks, Stones, & Shadows; Building the Egyptian Pyramids*. Norman, OK: Univ. of Oklahoma Press, 2001.
Jenkins, N. *The Boat Beneath the Pyramid*. New York: Holt-Rinehart-Winston, 1980.
Jequier, G. *Deux Pyramides du Moyen Empire*. Cairo: Service des Antiquites de l'Egypte, 1933.
Landstrom, B. *Die Schiffe der Pharaonen*. Haarlem: C. Bertelsmann Verlag, 1970.
Lauer, J.P. *La Pyramide a Degres*. 3 vols. Cairo: Imprimerie de l'Institut Francais, 1936-1939.
———. *Le Mystere des Pyramides*. Paris: Presses de la Cite, 1974.
———. *Saqqara*. London: Thames and Hudson Ltd, 1976.
Lehner, M. *The Pyramid Tomb of Hetep-heres and the Satellite Pyramid of Khufu*. Mainz: Philipp von Zabern, 1985.
———. *The Complete Pyramids*. London: Thames & Hudson, 1997.
———. and Z. Hawass. "Who Built the Pyramids?" *NOVA* interview 1997: http://www.pbs.org/wgbh/nova/ancient/who-built-the-pyramids.html

Lipke, P. *The Royal Ship of Cheops*. Oxford. BAR International Series 225, 1984.
Lucas, A. and J.R. Harris *Ancient Egyptian Materials and Industries*. Dover Edition, unabridged replication of fourth edition. London: Edward Arnold Ltd, 1962.
Mace, A.C. "Bulletin 9-Excavations at the north pyramid of Lisht." New York: Metropolitan Museum of Art, 1914.
Maragioglio, V. and C. Rinaldi. *L'Architettura della Piramidi Menfite*. 7 vols. Turin and Rapallo: Tipografia Canessa, 1963-1977.
Mendelssohn, K. *The Riddle of the Pyramids*. New York: Praeger Publishers, 1974.
Moores, R. "Evidence for Use of a Stone-Cutting Drag Saw by the Fourth Dynasty Egyptians." *JARCE*, volume 28, 1991.
Morell, V. "The Pyramid Builders," *National Geographic*, Nov. 2001.
Naville, E. *The Temple of Deir el Bahari*. Part VI. London: The Egypt Exploration Fund, 1908.
Neuburger, A. *The Technical Arts and Sciences of the Ancients*. New York: Barnes and Noble, 1969.
Newberry, P. *El Bersheh. Part 1. The Tomb of Tehuti-hetep*. London: Archaeological Survey of Egypt, 1893.
Nour, M., M.S. Osman, Z. Iskander, and A. Y. Moustafa. *The Cheops Boats*. Cairo: General Organisation for Government Printing Offices, 1960.
Petrie, W.M.F. *The Pyramids and Temples of Gizeh*. 2nd Ed. London: Field & Tuer, 1885.
———. *A History of Egypt*. London: Methuen & Co., 1895.
———. *Tools and Weapons*. London: British School of Archaeology in Egypt, 1917.
———. *Egyptian Architecture*. London: British School of Archaeology in Egypt, 1938.
———. *Wisdom of the Egyptians*. London: British School of Archaeology in Egypt, 1940.
Petrie, W.M.F., E. Mackay, and G. Wainwright. *Meydum and Memphis (III)*. London: British School of Archaeology in Egypt, 1910.
———. *The Labyrinth, Gerzeh, and Mazghuneh*. London: British School of Archaeology in Egypt, 1912.
Petrie, W.M.F., G. Brunton, and M. Murray. *Lahun II*. London: British School of Archaeology in Egypt, 1923.
Pochan, A. *The Mysteries of the Great Pyramids*. New York: Avon Books, 1971.
Protzen, Jean-Pierre. *Inca Architecture and Construction at Ollantaytambo*. New York & Oxford:
Oxford University Press, 1993.
Quibell, J.E. *Excavations at Saqqara*. Vols 1 & 2. Cairo: Imprimerie de l'Institut Francais, 1907, 1908.

Reisner, G.A. *Mycerinus*. Cambridge, MA: Harvard University Press, 1931.

———. *The Development of the Egyptian Tomb Down to the Time of Cheops*. Cambridge, MA: Harvard University Press, 1936.

———. *A History of the Giza Necropolis*. Vol 1. London: Harvard University Press, 1942.

Reisner, G.A. and W.S. Smith, *A History of the Giza Necropolis*. Vol 2. Cambridge, MA: Harvard University Press, 1955.

Robert, M.A. "Discoveries on the Summit of the Meidoum Pyramid." *Annales du Service des Antiquites de l'Egypte*, no. 3, 1902.

Romer, J. *The Great Pyramid*. Cambridge: Cambridge University Press, 2007.

Schoch, R.M. and R.A. McNally. *Pyramid Quest*. New York: Tarcher/Penguin, 2005.

Smith, C. B. *How The Great Pyramid Was Built*. Washington: Smithsonian Books, 2004.

Smith, W.S. *Our Inheritance in the Great Pyramid*. 3rd ed. London. Daldy, Isbister & Co., 1877.

Stadelmann, R. *Die agyptischen Pyramiden*. Darmstadt:Verlag Philipp von Zabern, 1985.

Strabo, *Geography, Book 17*, Translation by Horace L. Jones. Cambridge, MA and London: Harvard University Press, 1949.

Tellefsen, O. "A New Theory of Pyramid Building." *Natural History*. vol. 79, No. 9, 1970.

Tompkins, P. *Secrets of the Great Pyramid*. New York: Harper & Row, 1971.

Vyse, R.W.H. *Operations Carried on at the Pyramids of Gizeh*. Vol 1. London: James Fraser, 1840.

——— *Operations Carried on at the Pyramids of Gizeh*. Vol 3. London: Weale & Nickisson, 1842.

Printed in the United States
By Bookmasters